MODELLING AND ANALYSIS OF FINE SEDIMENT TRANSPORT

IN WAVE-CURRENT BOTTOM BOUNDARY LAYER

MODELLING AND ANALYSIS OF FINE SEDIMENT TRANSPORT IN

WAVE-CURRENT BOTTOM BOUNDARY LAYER

DISSERTATION

Submitted in fulfillment of the requirements of
the Board for Doctorates of Delft University of Technology
and
of the Academic Board of the IHE Delft
Institute for Water Education
for
the Degree of DOCTOR
to be defended in public on
Monday, 4 June 2018, at 15:00 hours
in Delft, the Netherlands

by

Liqin ZUO

Master of Engineering in Harbor, Coastal and Offshore Engineering, Nanjing Hydraulic Research
Institute, China
born in Shandong, China

This dissertation has been approved by the promotors:
Prof. dr. ir. J.A. Roelvink and
Prof. dr. Y.J. Lu

Composition of the doctoral committee:

Chairman	Rector Magnificus TU Delft
Vice-Chairman	Rector IHE Delft
Prof. dr. ir. J.A. Roelvink	IHE Delft/TU Delft, promotor
Prof. dr. Y.J. Lu	Nanjing Hydraulic Research Institute, China, promotor

Independent members:

Prof.dr. P. Nielsen	The University of Queensland, Australia
Prof.dr. L.C. van Rijn	University of Utrecht
Prof.dr.ir. Z.B. Wang	TU Delft
Prof.dr.ir. J.C. Winterwerp	TU Delft
Prof. dr.ir. W.S.J. Uijttewaal	TU Delft, reserve member

This research was conducted under the auspices of the Graduate School for Socio-Economic and Natural Sciences of the Environment (SENSE)

CRC Press/Balkema is an imprint of the Taylor & Francis Group, an informa business

Published by:
CRC Press/Balkema
Schipholweg 107C, 2316 XC, Leiden, the Netherlands
Pub.NL@taylorandfrancis.com
www.crcpress.com – www.taylorandfrancis.com
ISBN 978-1-138-33468-7.

The work is financially supported by the Joint Research Project of The Netherlands Organisation for Scientific Research (NWO) - National Natural Science Foundation of China (NSFC) (Grant No. 51061130546) and NSFC (Grant No. 51520105014 and 51509160).

To my family

Abstract

The evolution and utilization of estuarine and coastal regions are greatly restricted by sediment problems. Inspired by the Caofeidian sea area in Bohai Bay, China, this study aims to better understand silty sediment transport under combined action of waves and currents, especially in the wave-current bottom boundary layer (BBL), and to improve our modelling approaches in predicting estuarine and coastal sediment transport.

Field observations were carried out in northwestern Caofeidian sea area of Bohai Bay and field data were collected on several other silt-dominated coasts. Analysis shows that silt-dominated sediments are sensitive to flow dynamics: the suspended sediment concentrations (SSCs) increase rapidly under strong flow dynamics (i.e., waves or strong tidal currents which can stir up sediments), and high concentrations cause heavy sudden back siltation in navigation channels. In the following, details of silty sediment transport are studied, focusing on the BBL and high concentration layer (HCL).

From laboratory experiments and theoretical analysis, an expression for sediment incipient motion is proposed for silt-sand sediment under combined wave and current conditions. The Shields number was revised by adding the cohesive force and additional static pressure, leading to an extended Shields curve.

To study the HCL, a process based 1DV model was developed for flow-sediment dynamics near the bed in combined wave-current conditions. Based on the physical processes, special approaches for sediment movement were introduced, including approaches for different bed forms (rippled bed and 'flat-bed'), hindered settling, stratification effects, mobile bed effects, reference concentration and critical shear stress. The HCL was simulated and sensitivity analysis was carried out by the 1DV model on factors that impact the sediment concentration in the HCL. The results show that the HCL is affected by both flow dynamics and bed forms; the thickness of the HCL is about twice the height of the wave boundary layer; bed forms determine the shape of the concentration profile near the bottom, and flow dynamics determine the magnitude. For finer sediment, stratification effects and mobile bed effects impact the sediment concentration greatly.

Finally, based on the 1DV model, the formulations of the mean sediment concentration profile of silty sediments were studied. By solving the time-averaged diffusion equation for SSC and considering the effects of bed forms, stratification and hindered settling, expressions for time-averaged SSC profile under wave conditions were proposed for silt and are applicable for sand as well. Subsequently, the depth-averaged sediment concentration was yielded by integrating the SSC profile under wave conditions.

In summary, this research unveils several fundamental aspects of silty sediment, i.e., criterion of the incipient motion, the SSC profiles in HCL and their time-averaged parameterization in wave-dominated conditions. A 1DV model was developed for fine sediment transport in the wave-current BBL. The developed approaches are expected to be applied in engineering practice and further simulation.

Samenvatting* (Abstract in Dutch)

De evolutie en het gebruik van estuariene en kustgebieden worden sterk beperkt door sedimentproblemen. Geïnspireerd door het zeegebied van Caofeidian in Bohai Bay, China, heeft deze studie tot doel silt sedimenttransport beter te begrijpen onder gecombineerde actie van golven en stromingen, vooral in de golf-stroom bodemgrenslaag (BBL), en onze modelleringsbenaderingen te verbeteren bij het voorspellen van sediment transport in estuaria en kusten.

Veldobservaties zijn uitgevoerd in het noordwestelijke Caofeidian-zeegebied van Bohai Bay en veldgegevens zijn verzameld op verschillende andere door silt gedomineerde kusten. Analyse toont aan dat silt-gedomineerde sedimenten gevoelig zijn voor stromingsdynamica: de gesuspendeerde sedimentconcentraties (SSC's) stijgen snel onder sterke stromingsdynamica (dwz golven of sterke getijstromen die sedimenten kunnen doen opwoelen), en hoge concentraties veroorzaken zware, plotselinge aanslibbing in navigatiekanalen. In het volgende worden details van siltig sedimenttransport bestudeerd, met de nadruk op de BBL en de hoge concentratielaag (HCL).

Uit laboratoriumexperimenten en theoretische analyse wordt een uitdrukking voor het begin van beweging voorgesteld voor silt-zand sediment onder gecombineerde golf-en stroomomstandigheden. Het Shields-getal is herzien door de cohesiekracht en extra statische druk toe te voegen, wat leidde tot een uitgebreide Shields-kromme.

Om de HCL te bestuderen is een op processen gebaseerd 1DV-model ontwikkeld voor de dynamiek van het sediment en sediment in de buurt van het bed in gecombineerde golf-stroomcondities. Op basis van de fysische processen zijn speciale benaderingen voor sedimentbeweging geïntroduceerd, waaronder benaderingen voor verschillende bodemvormen (geribbeld bed en vlakke bodem), gehinderde bezinking, gelaagdheidseffecten, mobiele bodem-effecten, referentieconcentratie en kritische schuifspanning. De HCL is gesimuleerd en gevoeligheidsanalyse is uitgevoerd met het 1DV-model op factoren die de sediment concentratie in de HCL beïnvloeden. De resultaten tonen aan dat de HCL wordt beïnvloed door zowel stromingsdynamiek als bodemvormen; de dikte van de HCL is ongeveer tweemaal de hoogte van de golfgrenslaag; bodemvormen bepalen de vorm van het concentratieprofiel nabij de bodem en de stromingsdynamiek bepaalt de grootte. Voor fijner sediment hebben stratificatie-effecten en mobiele bodem-effecten een grote invloed op de sedimentconcentratie.

Ten slotte zijn, op basis van het 1DV-model, de formuleringen van het gemiddelde sedimentconcentratieprofiel van siltachtige sedimenten bestudeerd. Door de tijdgemiddelde diffusievergelijking voor SSC op te lossen en rekening te houden met de effecten van bodemvormen, stratificatie en gehinderde bezinking, zijn uitdrukkingen voor fase-gemiddeld SSC profiel onder golfcondities voorgesteld voor slib, die ook toepasbaar zijn op zand. Vervolgens is de dieptegemiddelde sedimentconcentratie verkregen door het SSC-profiel onder golfcondities te integreren.

Samengevat onthult dit onderzoek verschillende fundamentele aspecten van siltig sediment, d.w.z. criterium van de begin van beweging, de SSC-profielen in HCL en hun fase-gemiddelde parameterinstelling in door golven gedomineerde condities. Een 1DV-model is ontwikkeld voor fijn sedimenttransport in de golf-stroom BBL. De ontwikkelde benaderingen zullen naar verwachting worden toegepast in de engineering praktijk en verdere simulaties.

Contents

Chapter 1

Introduction

This chapter briefly introduces the research background, objectives and research questions. The state-of-the-art of sediment transport and its modelling under combined action of waves and currents are reviewed. Firstly, the differences of sediment behaviour are reviewed as well as a brief introduction of the wave-current bottom boundary layer. Then, the numerical simulation for sediment transport under combinations of waves and currents are reviewed and several key approaches in sediment transport modelling are discussed. In the end, based on these discussions, the objectives and research questions as well as the outline of this thesis are proposed.

1.1. Background

Estuarine and coastal regions in the world are the centres of socio-economic activities. Protection and utilization of coastal areas have been studied extensively, for example storm surge defences, development of harbours and navigation channels, and reclamations. The evolution and utilization of estuarine and coastal regions are largely restricted by sediment issues. Waves and currents are the main dynamic driving forces of sediment transport in estuary and coastal areas. The measured data and experiences from practice show that sediment transport under combined action of waves and tidal currents is the main factor causing deposition in ports and navigation channels in silty and muddy coasts. For example, during a storm surge, a large amount of sediment on the sea bed will be stirred up by waves and then be transported rapidly by currents. In addition, significant deposition may occur in the nearby channels (especially new excavated channels in shallow water area) due to the weakening of dynamics close to the bed. In China, such sudden siltation is a common phenomenon during storm surges (Lu et al., 2009; Sun et al., 2010). Thus, the study of wave-current movement and sediment transport is of great significance in both academic research and engineering practice, which has received much attention from many scholars and engineers.

In particular, silty sediment is much more sensitive to wave-current interaction, as it is easily to be stirred up and resettled down. Silt-dominated coastal areas are widely found, such as the eastern and southwestern Bohai Bay, the Jiangsu coast in China (Figure 1-1) and the Semen Tuban port sea area in Indonesia. Meanwhile, silt is the prevailing sediment fraction in some rivers, such as in the Yellow River and Yangtze River in China (Te Slaa et al., 2015). Under strong wave conditions it can be stirred up in large volumes, moved by currents and deposited near infrastructure like harbours, waterways and intakes. Due to its special behaviour, this kind of sediment has drawn much attention from researchers in recent years, such as studies on the hindered settling (Te Slaa et al., 2015), sediment movement (Cao et al., 2003) and reference concentration (Yao et al., 2015).

The Caofeidian sea area is a silt-dominated coast, which is located in Bohai Bay, China. The coastal system is complex, consisting of barrier islands, shoals, lagoons, channels, and inlets under the action of strong coastal dynamic forces, such as tides, waves, and storm surges (Lu et al., 2009). The tidal inlet and the nearby coasts form a sediment sharing system. The tidal inlet has the function of interception, capturing, allocation and transfer of littoral sand drift. As waves have great effects on the shorelines and tidal inlet, sediment transport is significantly influenced by the combined action of waves and tidal currents. In particular, the strong waves can influence the sediment concentration significantly in the shoal area. There are deep channels seaward and big shoals landward, which make the Caofeidian sea area a natural site for building a large-sized port. There have been several large development projects in the Caofeidian coastal area, including reclamation of harbour area, sand dredging, and excavation of harbour basin and navigation channels. In the overall plan, up to 310 km² areas of shoals are reclaimed for construction of the harbour, which is one of the biggest single reclamation projects in the world. Thus, we could conclude that in the Caofeidian sea area, with its complicated sediment composition and mixed flow-dynamics, it is urgent to study the sediment transport in this kind of coastal system.

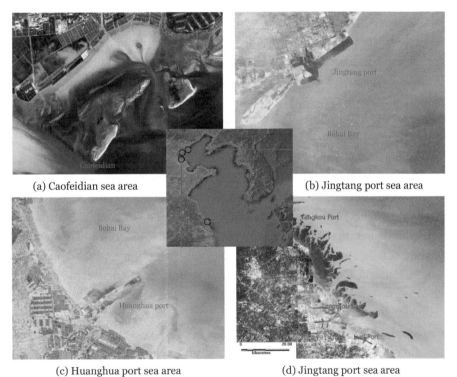

(a) Caofeidian sea area   (b) Jingtang port sea area

(c) Huanghua port sea area   (d) Jingtang port sea area

Figure 1-1. *Pictures of several silt-dominate coasts*

The interaction of wave-current-sediment occurs in the bottom boundary layer (BBL). To study the mechanisms of sediment transport, the BBL is an important and unavoidable field. Although the BBL is very thin, it is the initial place for shear stress, turbulence bursts, incipient motion and suspended sediment, thus it has great influence on sediment transport and morphology evolution. Many scholars presented treatments of sand sediment transport in wave-current conditions (Nielsen, 1992; van Rijn, 2007a). The details of fine sediment transport in the BBL still needs further study.

1.2. Behaviour of sediments with different grain sizes

Basically, distinguished by grain sizes, sediment can be classified into gravel, sand, silt, and clay (Table 1-1). Normally, sediment with grain size less than 62 μm (silt+clay) is defined as cohesive sediment and sediment with grain size larger than 62 μm is defined as non-cohesive sediment (Winterwerp and Van Kesteren, 2004). There is a fundamental difference in sedimentary behaviour between sand and clay materials (van Rijn, 1993). The reasons of different behaviour with different particle sizes are mainly their physical characteristics (e.g., inertial force), diffusion mechanism, flocculation etc. For larger particles (sand and gravel), sediments behave in a non-cohesive manner, for example, sediment particles consolidate instantaneously, the surface erodes particle by particle, and the bed load transport is the main type, etc. For smaller particles (clay), the sediments behave in a cohesive manner, for example, they consolidate relatively slowly, the surface erodes in aggregates (Righetti and Lucarelli, 2007), flocculation is a common phenomenon, and the suspended load transport is the main type.

Recent field observations and flume experiments have shown that silty sediment or silt-dominated sediment has a special behaviour, which is neither like typical sand (non-cohesive) nor like typical mud (cohesive). Erosion tests have suggested that silt-enriched mixtures exhibit cohesive-like behaviour (Roberts et al., 1998), but flocculation has not been observed based on settling experiments on silt (with clay contents less than 10%) (Te Slaa et al., 2015; Yao et al., 2015). Silt is often referred to as pseudo-cohesive or semi-cohesive sediment, to be differentiated from non-cohesive or cohesive materials. Silt may hold dual features of non-cohesive and cohesive sediments. It is natural that there is no clear separation of cohesive and non-cohesive sediments and it is reasonable to have a transition zone between them from a sense of continuity.

According to laboratory experiments in combination with field work in silt-rich environments (Te Slaa et al., 2013), the transitional behaviour in silt-rich sediment occurs at a threshold when the clay content is about 10%. Mehta and Lee (1994) suggested that the 10-20 μm size may be considered practically to be the dividing size that differentiates cohesive and cohesionless sediment behaviour. Stevens (1991) proposed 16 μm to be the division between sediments that flocculate significantly. Some experiments (Li, 2014; Yao et al., 2015; Zhou and Ju, 2007) showed that the grain size of 45 μm to 110 μm shared similar suspension behaviour under wave-current conditions. Some scholars defined the silty coast with medium grain size of 30 μm to 125 μm and the clay percentage less than 25%, to be differentiated with sandy coast and muddy coast (Cao et al., 2003). Thus, this study focuses on silt and very fine sand, defined as silty sediment, which is considered to be the transition zone of non-cohesive and cohesive sediments.

Table 1-1. Grain size model of American Geophysical Union (van Rijn, 1993)

Class Name	Millimeters	Micrometers	Phi Values
Boulders	>256		<-8
Cobbles	256-64		-8 to -6
Gravel	64-2		-6 to -1
Very coarse sand	2.0-1.0	2000-1000	-1 ~ 0
Coarse sand	1.0-0.5	1000-500	0 ~ +1
Medium sand	0.5-0.25	500-250	+1 ~ +2
Fine sand	0.25-0.125	250-125	+2 ~ +3
Very fine sand	0.125-0.062	125-62	+3 ~ +4
Coarse silt	0.062-0.031	62-31	+4 ~ +5
Medium silt	0.031-0.016	31-16	+5 ~ +6
Fine silt	0.016-0.008	16-8	+6 ~ +7
Very fine silt	0.008-0.004	8-4	+7 ~ +8
Coarse clay	0.004-0.002	4-2	+8 ~ +9
Medium clay	0.002-0.001	2-1	+9 ~ +10
Fine clay	0.001-0.0005	1-0.5	+10 ~ +11
Very fine clay	0.0005-0.00024	0.5-0.25	+11 ~ +12
Colloids	<0.00024	<0.024	>+12

The threshold behaviour of silty sediment is poorly understood (Mehta and Lee, 1994). The initiation of motion of non-cohesive sediments (sand and gravel) has been well studied with both experimental and theoretical works, such as the Shields curve. In contrast, relatively little experimental or theoretical work has been done on the initiation of motion of sediments consisting of cohesive particles (Lick et al., 2004). Since the Shields' curve is not very accurate for fine sediment beds, van Rijn (2007b) proposed empirical calibration factors (cohesive effects and packing effects) for fine sediments. Considering the cohesive force, Tang (1963) and Dou (2000) proposed a critical velocity for fine sediments. Lick et al. (2004) proposed a theoretical description for the fine sediment initiation of motion, including the cohesive forces between particles, as well as changes in bulk density. Righetti and Lucarelli (2007) proposed a threshold criterion for incipient motion of cohesive-adhesive sediments, which was also an extension of the Shields curve. As a transition zone of cohesive and non-cohesive sediment, the threshold of silt may be very complex. We still need to study critical shear stress under combined wave-current conditions for silt sediment from a unified perspective, for applying in modelling and engineering practice.

For hindered settling velocity, Richardson and Zaki (1954)'s formula is commonly used for sand, while Winterwerp and Van Kesteren (2004)'s formula is often used for cohesive sediment. Recently, Te Slaa et al. (2015) studied the hindered settling velocity of silt by experiments and proposed a formulation. According to their study, the largest difference between the settling of sand and silt-sized particles is the hydrodynamic regime in which these particles settle individually. The fluid movement around settling particles with d > 100 μm (sand) is turbulent, which causes these particles to settle outside the Stokes regime. Since the fluid movement around particles with d < 100 μm is laminar, silt particles settle in the Stokes regime and their geometry is not of influence on the hindered settling. A more generic hindered settling formulation for silt is derived by Te Slaa et al. (2015), in which no physical processes characteristic for cohesive sediment are included, indicating that the hindered settling of silt can by described by noncohesive processes. Under sheet flow conditions, near the bed level z=0 in high-concentration area (0.3<c<0.4), Nielsen et al. (2002) found that the settling velocity is much lower than expected on the basis of the formulae suggested by Richadson and Jeronimo (1979). It means that, when penetrate into the sheet flow layer, we have to carefully use the formulae of hindered settling velocity that normally come from settling column experiments.

A high concentration layer (HCL) normally exists near the bottom under wave-dominated conditions, which is one of the most important characteristics of silt and fine sand. Sediment suspension is limited by the high oscillatory motion, and the sediment concentration near the bottom is much higher than that in the upper part. The HCL has been found in laboratory experiments (Dohmen Janssen et al., 2001; Yao et al., 2015) and field observations (Te Slaa et al., 2013). Some literature (Kineke et al., 1996; Trowbridge and Kineke, 1994) defines high concentration at the elevation where the concentration c = 10 kg/m³, or c = 21 kg/m³ by Winterwerp (1999). Lamb and Parsons (2005) defined the thickness of the high concentrated mud layer as the elevation where the concentration $c = 0.1c_{bed}$ (where c_{bed} is arbitrarily set at 1,400 g/l). Experiments have shown that there is a distinct interface between the HCL near the bottom and the clear water in the upper part under wave conditions, as shown in Figure 1-2 (Lamb and Parsons, 2005) and Figure 1-3 (Yao et al., 2015). In this study the HCL is defined as the higher concentration layer below where the gradient of sediment concentration changes abruptly in the upper part.

It can be concluded that, though the sand and mud have been studied intensively, the silt sediment is still under researched, in particular on the incipient motion, the details of the HCL and the modelling approaches.

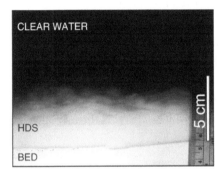

Figure 1-2. *The high concentration near the bottom measured in a flume experiment with median sediment grain size of 20-66 μ m (Lamb and Parsons, 2005)*

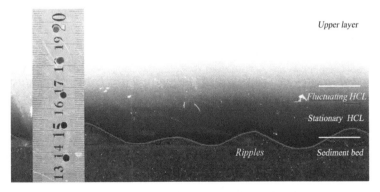

Figure 1-3. *High concentration layer (HCL) measured in a flume experiment with median sediment grain size of 88 μ m (Yao et al., 2015) (The darker colour represents higher sediment concentration)*

1.3. Wave-current bottom boundary layer (BBL)

The sediment transport under waves and currents is governed by the turbulence in BBL. Boundary layer is the thin layer near the solid wall where viscosity has great influence on viscous flow. For low viscosity flow such as water or air, viscosity only exists in the boundary layer near the wall surface.

For steady flow or tidal currents, the boundary layer could develop fully and be as thick as the whole water depth in shallow water areas. The velocity distribution then generally meets a logarithmic law of the wall. However, oscillatory flow moves fast and the wave boundary layer cannot fully develop in wave conditions because of the short period, and the thickness is only in the order of centimetres or even millimetres.

The measured velocity profiles in wave BBL show a velocity overshoot at a certain height from the bed (Figure 1-4). The overshoot is a special phenomenon in oscillatory and wave boundary layers. The velocity overshoot occurs because the velocity defect $u_\infty(t) - u(z,t)$ has the nature of a damped wave which alternately adds to and subtracts from the free stream velocity $u_\infty(t)$ (Nielsen, 1992), in which $u(z,t)$ is the orbital velocity above the bottom.

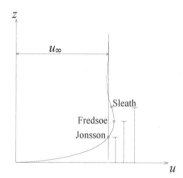

Figure 1-4. *Sketch of velocity distribution in wave bottom boundary layer (after Jonsson (1966) and You (1994))*

When waves and currents coexist, the boundary layer becomes more complicated. The study of interaction of waves and currents started from Unna (1942), who studied the wave field affected by tidal currents. Longuet-Higgins and Stewart (1962) proposed the radiation stress to explain wave-current interaction, which greatly accelerated the progress of this study. Many scholars studied the wave-current BBL flow structure by flume experiments (e.g., Bakker and Van Doorn, 1978; van Doorn, 1981; Klopman, 1994) or mathematical models (e.g., Grant et al., 1984; You et al., 1991; Zhang et al., 2011; Zhao et al., 2006).

In the wave-current BBL, the wave boundary layer is nested in the bottom, while the current boundary layer could often reach the whole water depth. Therefore, there are mainly two parts: the bottom wave control layer and the upper flow control layer. Current has little effects on the structure of wave boundary layer and pure wave BBL theory is still applicable in wave-current BBL. Beyond the wave BBL, the turbulence and shear stress are determined by currents. Figure 1-5 shows the difference of velocity distribution in the BBL of waves and currents. Waves have higher oscillatory frequency, the gradient of velocity is higher, and thus the shear stress is higher. The turbulence is restricted in the bed surface in wave conditions, while the turbulence could reach the whole water depth in current conditions. Waves could enhance the bottom shear stress and turbulence. Formation and development of BBL under wave-current interaction determines the magnitude of the bed shear stresses and velocity distribution near the bed.

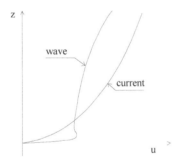

Figure 1-5. Distribution of velocity in wave and current boundary layers (after Nielsen (1992))

1.4. Numerical simulation for sediment transport under combined waves and currents

The numerical simulation of hydrodynamic and sediment transport processes is a powerful tool in the description and prediction of morphological changes and sediment budgets in the coastal zone (Da Silva et al., 2006). Sediment transport modelling under combined action of waves and currents started in last century with the development of simple analytical and 1DV models (Grant and Madsen, 1979; Grant et al., 1984; Smith and McLean, 1977; Stive and De Vriend, 1994). These models solve boundary layer equations to obtain several key variables' distribution, and are frequently used for theoretical study. Because of its simplicity and precision, it is valuable for some special issues, such as intra-wave vertical distribution of velocity, shear stress, concentration etc. However, the 1DV model could not resolve the horizontal patterns. Then, complex area models are often used, which solve 2D or 3D conservation equations of wave, current and sediment transport.

1.4.1. 1DV model for flow-sediment movement in wave-current BBL

1DV models employ the Reynolds equations derived from the N-S equations in x-z coordinates. How to determine the eddy viscosity υ_t is a key issue in 1DV models. The methods mainly include time-invariant eddy viscosity model, mixing length model, turbulent models, etc. Most early models used linearized boundary layer equations and time invariant algebraic eddy viscosity (e.g., Christoffersen and Jonsson, 1985; Grant and Madsen, 1979; Myrhaug and Slaattelid, 1990; Sleath, 1995; Smith and McLean, 1977; You, 1994). Some scholars also present development of the time-variant algebraic eddy viscosity (Madsen and Wikramanayake, 1991; Trowbridge and Madsen, 1984). The mixing length model was also widely employed in early models (Bakker and Van Doorn, 1978; Bijker, 1967; O'Connor and Yoo, 1988; Yang et al., 2006). In the more refined modelling of the BBL, the focus has been put on the turbulence closure, which led to the development of several one-equation models (Davies et al., 1988; Madsen, 1994) and two-equation models (Henderson et al., 2004; Holmedal and Myrhaug, 2009; Kranenburg et al., 2013; Li and Davies, 1996); in particular, the k-ε model was widely used (Holmedal and Myrhaug, 2009; Kranenburg et al., 2013; Zhang et al., 2011).

In the past, different models were developed to predict the sediment transport under waves or wave-current conditions. These models could be divided into three different classes (Hassan and Ribberink, 2010; Zhang et al., 2011): empirical quasi-steady transport models, intermediate transport models and full unsteady sediment transport models (process based). Quasi-steady transport models assumed that the intra-wave sediment transport reacts immediately to the time-dependent horizontal flow velocity throughout the wave cycle, without any phase difference between the flow velocity and the concentration (Grant and Madsen, 1979; Nielsen and Callaghan, 2003; Ribberink and Al-Salem, 1995; Sleath, 1978). Intermediate transport models are also empirical transport formulas, but phase-lag effects are included in a parameterized way (Ahmed and Sato, 2003; Camenen and Larson, 2005; Dibajnia and Watanabe, 1992; Dohmen-Janssen et al., 2002). Full unsteady sediment transport models are based on a full time-dependent simulation of both velocities and concentrations during the wave cycle at different elevations above the bed (Fredsøe, 1984; Guizien et al., 2003; Hassan and Ribberink,

2010; Holmedal and Myrhaug, 2006; Holmedal and Myrhaug, 2009; Kranenburg et al., 2013; Ribberink and Al-Salem, 1995; Ruessink et al., 2009; Uittenbogaard et al., 2001). The process-based unsteady models are advanced approaches.

Recently, instead of sinusoidal waves, many models focus on the effects of non-linear wave characteristics on sediment transport, such as wave asymmetry (Ruessink et al., 2009), progressive wave streaming (Rodrıguez, 2009), effects of the free surface on the BBL (Kranenburg et al., 2013), acceleration skewness (King, 1991; Nielsen, 1992), phase-lag effects, ripple bed (Davies and Thorne, 2005) and sheet flow (Kranenburg et al., 2013). These approaches assist in exploring the small-scale near-bed sediment dynamics (Davies et al., 2002).

If the sediment concentration is high, sediment-induced turbulence damping can largely affect velocity profiles and transport rates, especially for fine sediment (Conley et al., 2008; Hassan and Ribberink, 2010; Kranenburg et al., 2013; Winterwerp, 2001). Generally, the buoyancy flux B_k accounts for the conversion of turbulent kinetic energy to mean potential energy with the mixing of sediment, treated equivalent to buoyancy flux in a salt-stratified or thermally stratified flow.

Up to now, many models focus on sand (e.g., Dong et al., 2013; Uittenbogaard et al., 2000) and fluid mud (e.g., Hsu et al., 2009; Winterwerp and Uittenbogaard, 1997), but few on silty sediments. So far, there is still little thorough modelling and parameterization of sediment concentration distribution in high concentration layer (HCL) of silty sediments.

As far as we know, the eddy coefficient, velocity and shear stress distribution in wave BBL and current BBL are different. In combined wave-current BBL, there are wave control layer in bottom and current control layer in upper part, which cause complex uneven sediment concentration profiles in different combination of waves and currents. Based on verification of experiment data, the 1DV model is a good tool to supply more details of turbulence, diffusion coefficient, inflection point of high sediment concentration etc., which is helpful for us to explore the characteristics of HCL. Thus, it is still urgent to study the HCL modelling under combination of waves and currents.

1.4.2. Coastal area models: 2DH and 3D models

For different study areas, three types of models have been classified (Roelvink and Reniers, 2012): coastal profile models, where the focus is on the cross-shore processed and the long-shore variability is neglected (Roelvink and Brøker, 1993; Schoonees and Theron, 1995), coastline models, where the cross-shore profiles are assumed to retain their shape even when the coast advances or retreats (Szmytkiewicz et al., 2000) and coastal area models, where variations in both horizontal dimensions are resolved (Nicholson et al., 1997).

Coastal morphological area models had been developed since the early 1980s (De Vriend et al., 1993; Nicholson et al., 1997). So far, there are some robust and flexible models, such as Delft3D (Deltares, 2014), XBeach (Roelvink et al., 2009; Roelvink et al., 2018), Mike (Pietrzak et al., 2002), Telemac (Villaret, 2010; Villaret et al., 2013), ECOMSED (Blumberg, 2002), ADCIRC (Luettich and Westerink, 2004), ROMS (Warner et al., 2008), Wallingford, COHERENS (Luyten et al., 2006), etc. In this paper, we will not describe these models; instead, we will try to list some key problems and approaches of the wave-current-sediment simulation.

Based on model's theory, morphological area models can generally be classified into process-based or behaviour-based models (Amoudry and Souza, 2011). The first approach is based on representing all relevant sediment transport processes. The second approach implements simple parameterized descriptions of the general behaviour of the morphological system at the larger scales of interest (centennial to geological) and relies essentially on long-term data sets for calibration.

Here we consider the process-based coastal area models which are being increasingly used to study coastal sediment dynamics and coastal morphology (Amoudry and Souza, 2011). These coastal area models are further subdivided into two-dimensional horizontal (2DH) models (e.g., Dissanayake et al., 2012; Ferrarin et al., 2008; Kuang et al., 2012), quasi-3D models (e.g., Drønen and Deigaard, 2007; Li et al., 2007) and three-dimensional (3D) models (e.g., Lesser et al., 2004; Liang et al., 2007; Normant, 2000; Pandoe and Edge, 2004; Pietrzak et al., 2002; Pinto et al., 2012; Wai et al., 2004; Warner et al., 2008).

Sediment transport is a complex, multidimensional, and dynamic process that results from the interactions of hydrodynamics, turbulence, and sediment particles. Grains can be transported by currents (tide driven, density driven, wave driven, or wind driven), wave motions, and combinations of the two. It is customary to keep a distinction between bed load and suspended load as they correspond to different physical mechanisms.

(1) Suspended load sediment transport

There are two methods for suspended load transport simulation: sediment transport rate (combining transport rate of waves and currents) and diffusion-advection equation. The latter one is widely used.

Following the Reynolds average method, we can obtain suspended load transport equations under combined action of waves and currents by dividing instantaneous sediment concentration into averaged part, wave related part and pulsation part. The equation forms are similar to the current-related sediment transport equations. Wave's effects must be considered in the source/sink terms, boundary conditions and turbulent diffusion coefficient.

3D suspended load sediment transport equation:

$$\frac{\partial c}{\partial t}+u\frac{\partial c}{\partial x}+v\frac{\partial c}{\partial y}+(w-\omega_s)\frac{\partial c}{\partial z}=\frac{\partial}{\partial x}(\varepsilon_x\frac{\partial c}{\partial x})+\frac{\partial}{\partial y}(\varepsilon_y\frac{\partial c}{\partial y})+\frac{\partial}{\partial z}(\varepsilon_z\frac{\partial c}{\partial z}) \tag{1-1}$$

2DH depth-averaged suspended load sediment transport equation:

$$\frac{\partial h\overline{c}_h}{\partial t}+\overline{u}_h\frac{\partial h\overline{c}_h}{\partial x}+\overline{v}_h\frac{\partial h\overline{c}_h}{\partial y}=\frac{\partial}{\partial x}(\varepsilon_x\frac{\partial h\overline{c}_h}{\partial x})+\frac{\partial}{\partial y}(\varepsilon_y\frac{\partial h\overline{c}_h}{\partial y})+F_s \tag{1-2}$$

where h is water depth, c is suspended load sediment concentration, u, v, w are velocities in x, y, z direction, $\overline{u}_h, \overline{v}_h, \overline{c}_h$ are depth-averaged value, $\varepsilon_x, \varepsilon_y, \varepsilon_z$ are turbulent diffusion coefficients, and F_s is the source/sink term.

(2) Bed load or total load sediment transport

There are also two methods for bed load transport simulation: sediment transport rate (combining transport rate of waves and currents) and diffusion-advection equations. The first one is most used now.

In early models, the bed-load transport formulas for current-only were adapted to combined current-wave situations by adapting the dimensionless shear stress (e.g., Einstein, 1950; Meyer-Peter and Müller, 1948). After that, most researchers have resorted to developing formulations directly fitted against as many datasets as they could get hold of (Gonzalez-Rodriguez and Madsen, 2007; Ribberink, 1998; Soulsby, 1997).

Similar with suspended load sediment, some scholars derived the diffusion-advection equation for bed load, such as Dou (2001). Wu et al. (2010) obtained total-load sediment transport equations by combing the suspended-load and bed load sediment transport model.

1.5. Key problems in sediment transport modelling

Numerical models are dependent on the theory of the flow dynamics and sediment transport. For a good modeller, we should have deep knowledge on sediment transport theory. How to improve the sediment transport model? As the mathematical models have been developed for years, the discretization schemes and calculation methods are already advanced for present study to some extent, though the endeavour on efficient computing is still ongoing. Thus, we do not focus on the computing method in this study; instead, we focus on the mechanisms of sediment transport.

The approaches in numerical models mainly depend on the understanding of the sediment transport near the bottom, such as the bed roughness, incipient critical shear stress, vertical distribution of sediment diffusivity and SSC, reference concentration, sediment source/sink terms etc. These processes directly or indirectly relate to the sediment transport in BBL.

1.5.1. Bed forms

Bed forms and bed roughness directly affect the bed shear stress, flow structure, and sediment concentration near the bed. Thus, an accurate simulation of the sediment alluvial process in boundary layer requires a delicate understanding of bed forms and their related roughness.

It is found that during the early dynamic-increasing period, ripples occur with height in the range of several centimetres and length in the range of tens of centimetres. When dynamics become stronger, sand dunes appear with height in the range of tens of centimetres and length in the range of hundreds; with further strengthened dynamics, the bed becomes smooth, exhibiting sheet flow (Li and Amos, 1999). Over rippled beds, the boundary layer separates behind the crests and vortex formation and shedding occurs during each wave half cycle. This phenomenon gives rise to a fundamentally different spatial and temporal distribution of momentum transfer in the near seabed layer compared with that above a flat bed.

The feedback interactions between the hydrodynamics, bed forms and sediment properties were investigated by some researchers (Hooshmand et al., 2015; Lofquist, 1986; Ribberink et al., 2007; Soulsby, 1997; Thorne and Hanes, 2002). The presence of bed forms modifies the hydraulic roughness, bottom stress, near-bed turbulence and sediment entrainment; these processes in turn induce different bed-form patterns. Based on the measured wave energy dissipation from Carstens et al. (1969), Nielsen (1992) argued that the bed roughness under oscillatory sheet-flow is of the order $100d_{50}$ or more, while the rippled beds the roughness is generally in the range $[100d_{50} - 1000d_{50}]$. The shape of the concentration profile will depend strongly on the bed form geometry (Nielsen, 1995). Over vortex rippled bed, sediment suspension near the bottom is dominated by organized vortex, which enhances the separation of the flow and the production of turbulence (Sato et al., 1985). The vortices are highly effective in transporting sediment to far greater heights above a rippled bed than occurs above plane beds (Bijker et al., 1976; Davies et al., 2002). While in sheet flow regime, the sediment concentration is dominated by random turbulence. Thus, the effects of bed forms have to be considered.

For silt, the effects of bed forms are much more important since the bed forms transform easily. Normally, the criterion of bed forms can be represented by mobility number $\psi = u_{wc}^2 / [(s-1)gd_{50}]$, where u_{wc} = the velocity of combined wave-current, s = 2.65 = relative density, g = gravity acceleration, and d_{50} = median grain size. According to O'Donoghue et al. (2006), flat bed (sheet-flow) regime prevails when $\psi > 300$, the ripple regime happens when $\psi < 190$ and a transition regime prevails when $190 < \psi < 300$. From Figure 1-6, it can be seen that, while sheet flow only exists in strong dynamics condition for sand (when $u_m > 1.0$ m/s for d_{50} = 200 μm), silt may experience both rippled bed and sheet flow under common conditions (when u_m = 0.30-0.38 m/s for d_{50} = 30 μm).

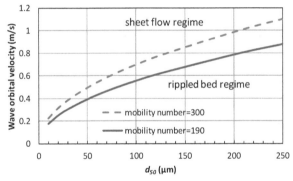

Figure 1-6. *The criterion conditions of bed forms according to* O'Donoghue et al. (2006)

1.5.2. Cohesive and non-cohesive sediment modelling

In many models, the simulation methods of sand transport and mud transport are treated separately. Special approaches were taken into account in aspect of incipient motion, settling velocity, deposition rate, erosion rate etc. (Ye, 2006).

Because of the relatively high settling velocities of sand grains, the transport of sand adjusts very quickly to hydrodynamic variations. Thus, empirical formulae of horizontal fluxes that are generally validated under equilibrium conditions can be used to model sand transport. These formulae can describe total (bed load + suspended load) sand transport or only the bed load fraction. For example, the methods of Brown (1950), van Rijn (1993) and Yalin (1963) are widely used to predict bed load transport; the methods of Bagnold (1963) and Engelund and Hansen (1972) are often used to predict the total load transport. On the other hand, cohesive sediments are mainly transported in suspension and are calculated by solving the advection/diffusion equation.

The methods for bed roughness are different for cohesive and non-cohesive sediment. For sand, the bed form roughness and the grain roughness ($2d_{50}$ or $2.5d_{50}$) (e.g., Li and Amos, 2001; Nielsen, 1992; van Rijn, 2007b) are often included in the model to provide bed roughness prediction. For cohesive sediment, a default friction factor and a default bed roughness are often defined (Ferrarin et al., 2008; Soulsby, 1997). Effects of bed roughness on boundary layer parameters are included in the computation of friction factor and effective bed shear stress.

For cohesive sediment, some models consider the consolidation (Normant, 2000; Villaret and Latteux, 1992). Self-weight consolidation had been modelled using a simplified, empirical numerical model (Neumeier et al., 2008).

Different approaches have been used to compute the net sediment flux between the water column and the bottom (the benthic flux) for cohesive and non-cohesive sediments. For non-cohesive sediments, the net sediment flux between the bottom and the water column is computed as the difference between the equilibrium concentration and the existing concentration in the lower level (Lesser et al., 2004). The resuspension and deposition of cohesive sediment were parameterized by several formulas (Ariathurai and Krone, 1976; Dou, 2001; Liu and Yu, 1995; Parchure and Mehta, 1985), while some formulations (e.g., McLean, 1992; Van Rijn, 1993) were adopted for non-cohesive sediment fractions.

At present, silt is normally categorized to cohesive sediment modelling. However, as mentioned in section 1.2, silt belongs to the transition zone between sand and mud and shows special behaviour; though there are some approaches developed for simulating the silty sediment movement (Liu, 2009; Te Slaa et al., 2015; Yao et al., 2015), the modelling approaches for silt sediment are still under-researched.

1.5.3. The key problems in 1DV case

To simulate the details of flow-sediment mechanisms is a big challenge. 1DV models focus on the bottom boundary layer, which are helpful to understand the vertical process of sediment transport. These make the models preferable for exploring the small-scale near-bed sediment dynamics (Davies et al., 2002). The key problems in 1DV model are the approaches of flow-sediment dynamics in BBL. These approaches that relate to sediment transport include the sediment diffusivity, reference concentration, pick-up function, hindered settling velocity, flow-sediment interactions etc. The precise of the model depends greatly on the understanding of mechanism process and is an ongoing research topic.

The boundary condition of sediment concentration at a reference level z_a is normally described as an upward pick-up flux and a downward settling flux. For suspended sediment, many models employ the time-depended vertical gradient of the near-bed sediment concentration as the pick-up flux at a reference level z_a :

$$\varepsilon_s \frac{\partial c}{\partial z} = -\omega_s c_a \qquad \text{at } z=z_a \qquad (1\text{-}3)$$

where c_a is the near-bed reference concentration. The choice of c_a is important. For sand simulation in sheet flow condition and rippled beds, Zyserman and Fredsøe (1994)'s formula and Nielsen (1992)'s formula were recommended respectively. Yao et al. (2015) proposed a formulation for silt sediment based on van Rijn's formula. However, gradient diffusion is not the appropriate conceptual framework for the domain close to the sheet flow layer (Nielsen, 2002). Nielsen (2002) found that the gradient diffusion model cannot simulate the measured phenomenon, such as sediment concentrations are almost constant near the undisturbed bed level. Then, he proposed the total sediment flux Q_z as a composition of a pick-up function $p(t)$ near $z=0$ and a settling flux,

$$Q_z(z,t) = p(t) - \omega_s c \qquad \text{at } z=0 \qquad (1\text{-}4)$$

At present, the pick-up function formulas are most for sand and few for fine sediment (e.g., Nielsen, 2002; van Rijn, 1984a). Furthermore, as the suspended sediment contributes to the main part of fine sediment transport, we focus on the upper suspension layer and do not penetrate into the sheet flow layer, thus the gradient diffusion method is still employed.

The sediment suspension mechanisms are different between rippled bed and plane bed, i.e., the maximum c_a happens nearly at the phase of maximum flow shear dynamics under plan bed conditions; while over rippled bed, it happens at the time of flow reversal because of the effects of the vortex. Some scholars, e.g., Davies and Thorne (2005) and Nielsen (1992) studied the pick-up function for rippled bed.

Hindered settling is another important parameter, i.e., the settling velocity will be reduced when the sediment concentration gradient is high. Sand, silt and mud have different behaviour of hindered settling (Te Slaa et al., 2015; Winterwerp and Van Kesteren, 2004). Richardson and Zaki (1954), Te Slaa et al. (2005) and Winterwerp (2004) proposed the formulation of hindered settling for sand, silt and mud, respectively.

The vertical gradient of sediment concentration would affect the turbulence, the so-called stratification effects (Winterwerp, 1999). Generally, the buoyancy flux B_k accounts for the conversion of turbulent kinetic energy to mean potential energy with the mixing of sediment. van Rijn (2007a) proposed a damping coefficient ϕ_d to evaluate the damping of vortex viscosity. Conley et al. (2008) found that, the effects of sediment stratification scale with orbital velocity divided by sediment setting velocity. By comparing with their filed data, Traykovski et al. (2007) tested the stratification effects on sediment concentration profile of fine sediment by a 1DV model, and the results showed that the stratification effects is a non-neglected term.

1.5.4. The source/sink term in 2DH case

The source/sink term F_s is a key issue in 2DH models, defined as $F_s = E\text{-}D$, with E is the erosion rate and D is the deposition rate. Different methods are used for the source/sink term, including the erosion and deposition flux method and the sediment transport capacity (equilibrium concentration) method.

i) Erosion and deposition flux method: Pick-up rate (upward sediment flux) and deposition rate (downward sediment flux)

The erosion and deposition flux method provides formulas directly relating the erosion flux to the flow (shear stress) and sediment parameters. This is frequently used for cohesive sediment, and also has recently been extended to non-cohesive sediment (Warner et al., 2008).

One of the most used formulas of this type is:

The erosion rate (Partheniades, 1965):

$$E = M\left(\frac{\tau_b}{\tau_e} - 1\right) \qquad for \qquad \tau_b > \tau_e \tag{1-5}$$

The deposition rate(Ariathurai and Krone, 1976; Krone, 1962):

$$D = \alpha_s \omega c\left(1 - \frac{\tau_b}{\tau_d}\right) \qquad for \qquad \tau_b < \tau_d \tag{1-6}$$

in which τ_b is bed shear stress, τ_e is erosion critical shear stress, τ_d is deposition critical shear stress, M is erosion coefficient, α_s is the sediment settling probability, and c is the sediment concentration.

More expressions could also be found in Ariathurai and Arulanandan (1978), Ariathurai and Krone (1976), Lumborg and Windelin (2003), Parchure and Mehta (1985), Waeles et al. (2007) and Wang and Pinardi (2002).

ii) Sediment transport capacity (equilibrium concentration) method

If we adopt the entrainment rate E & D as following (van Rijn, 1987),

$$E = \omega_s c_*$$
$$D = \omega_s c$$
(1-7)

Considering the recovery saturation coefficient α_r, thus, the source/sink term is

$$F_s = \alpha_r \omega_s (c_* - c)$$
(1-8)

where c_* is the depth-averaged sediment transport capacity (equilibrium concentration).

Galappatti and Vreugdenhil (1985) derived another expression:

$$F_s = \frac{h(c_* - c)}{T_s}$$
(1-9)

where T_s is a typical timescale.

This method was widely used for non-cohesive sediment (Amoudry and Souza, 2011). From Dou et al. (1995) and Liu (2009)'s approach, it can also be used for fine sediments modelling.

1.5.5. The source/sink term and bottom boundary conditions in 3D case

In the 3D case, the sediment enters the model through the bed boundary conditions and is transported further by the advection-diffusion equation, using the turbulence structure from the flow model or from empirical formulation for the eddy diffusivity distribution.

The exchange of sediment with the bed is implemented by way of sediment sources and sinks placed near the bottom computational cell (Lesser et al., 2004; Pinto et al., 2012; Warner et al., 2008). The boundary conditions must be satisfied at the bottom and the interface between the bed-load and the suspended load. There are usually two approaches to specify the suspended-load bottom boundary (Wu, 2008). One approach is to assume the near-bed suspended-load concentration to be at equilibrium.

$$c_{bottom} = c_{a*}$$
(1-10)

where c_{a*} is the equilibrium sediment concentration at the interface.

Another approach is that the flux of sediment between the bed and the flow can be approximated by:

$$(-\varepsilon_s \frac{\partial c}{\partial z} - \omega_s c)\big|_{bottom} = E - D \tag{1-11}$$

in which $E\text{-}D$ is the net erosion minus deposition flux of sediment. Positive value means occurrence of erosion, while negative value means occurrence of deposition.

The first approach is often called concentration boundary conditions, and the second approach is often called gradient boundary conditions. The first method is applicable for equilibrium sediment transport, while the second one is applicable for both equilibrium and non-equilibrium sediment transport (Wu, 2008). The second method is more general, and is employed by most models.

As mentioned above, there are mainly two ways to calculate erosion and deposition flux. The first one is to establish shear stress with erosion and deposition flux (Ariathurai and Arulanandan, 1978; Ariathurai and Krone, 1976; Lumborg and Windelin, 2003; Parchure and Mehta, 1985; Waeles et al., 2007; Wang and Pinardi, 2002). The second one is the method of comparison between the reference concentration and the actual concentration (Dou et al., 1995; Galappatti and Vreugdenhil, 1985; Liu, 2009; van Rijn, 1987). In many models, the reference concept concentration c_a is often used to calculate the source term $E = \omega_s c_a$ and $D = \omega_s c$.

However, the source and sink terms obtained above is near the bottom, at a reference height z_a above the bottom, and it has to be transferred from the bottom to the centre of the lowest computational cell. It depends on the assumption of the concentration distribution near bed. Lesser et al. (2004) assumed a linear concentration gradient between the calculated reference concentration at z_a and the computed concentration in the reference cell. The resulting expressions are:

$$E = c_a \left(\frac{D_v}{\Delta z} \right)$$
$$\tag{1-12}$$
$$D = c_{kmx} \left(\frac{D_v}{\Delta z} + w_s \right)$$

in which D_v is the vertical diffusion coefficient at the bottom of the reference cell, Δz is the vertical distance from the reference level z_a to the centre of reference cell, and c_{kmx} is the mass concentration of the sediment fraction in the reference cell (solved implicitly).

Pinto et al. (2012) and Villaret (2010) assumed the Rouse type of the sediment concentration profile between the reference height and the bottom cell centre. The resulting expression of E is:

$$E \approx E_a (\frac{a}{\Delta z})^\alpha \tag{1-13}$$

where E_a is the erosion flux at the reference height z_a, $\alpha = \omega_s / (\kappa u_*)$ is Rouse number, κ is Karman number and u_* is shear velocity.

These approaches make the erosion flux independent of the vertical resolution near the bottom, thereby eliminating the need to tune the erosive flux when the vertical grid is modified.

1.6. Objectives and research questions

The overall aim of this study is to better understand silty sediment transport under combined action of waves and currents, especially in wave-current BBL, and to improve our modelling approaches in predicting estuarine and coastal sediment transport. From the literature review, this study focuses on silt and very fine sand, which is considered to be the transition zone of non-cohesive and cohesive sediments. Thus, we have to have a unified perspective for fine-coarse sediment to study this kind of sediment. Some key approaches for sediment transport modelling are expected to be studied, such as the threshold motion, the SSC (suspended sediment concentration) profile, the equilibrium or saturation concentration near the bottom and the depth-averaged SSC. A 1DV model for fine sediment transport in wave-current BBL is expected to be developed and used to study the details of flow-sediment dynamics near the bottom. This study is a fundamental sediment research topic. The details of objectives are:

(1) An approach of critical shear stress of sediment incipient motion for silty sediment under wave, current and combined wave-current conditions.

(2) Combined with the experimental datasets, a process based 1DV model for fine sediment will be developed to study the behaviour of SSC in wave-current BBL. It is expected to provide a supplement of experimental data to study flow-sediment movement in BBL.

(3) The details and impact factors of HCL (high concentration layer) in wave-dominate conditions are expected to be revealed.

(4) Based on experimental data, field data and the 1DV model, parameterization of time-averaged SSC profile and depth-averaged SSC are expected to be studied, expecting to improve the 2DH and 3D simulation approaches for silty sediment.

The research questions are as follows:

(1) How to develop the threshold criterion for silt, considering the

differences and similarities between waves and currents, coarse and fine sediment?

(2) What are the special key approaches for fine sediment modelling in BBL?

(3) What's the relationship between HCL and BBL? What are the main impact factors of SSC profile in HCL?

(4) How to develop time-averaged approaches for fine sediment modelling from BBL and HCL study for practical purposes (e.g., to improve the approaches of sediment simulation in 2DH/3D model)?

1.7. Thesis organization

This thesis in all has six chapters (Figure 1-7):

(1) Chapter one, this introduction, presents the research background and overviews of sediment behaviour with different grain sizes and sediment transport modelling under combined action of waves and currents. Objectives and research questions are proposed.

(2) Chapter two describes the field observation on sediment concentration under waves and currents in northwestern Caofeidian sea area, as well as data collection in some silt-dominated coasts. General features of silt-dominated sediment movement are summarized.

(3) Chapter three is about incipient motion of silt-sand under combined waves and currents and an expression was proposed.

(4) Chapter four develops a 1DV model for flow-sediment simulation in wave-current BBL. Some key processes that were included in the model are represented through approaches for different bed forms (rippled bed and 'flat bed'), hindered settling, stratification effects, reference concentration and critical shear stress. Discussions are made on some factors which would impact the SSC profile of the HCL by the 1DV model.

(5) Chapter five parameterizes the mean SSC profile and depth-averaged sediment concentration for silty sediment.

(6) Chapter six summarizes the whole work and makes suggestions for future study.

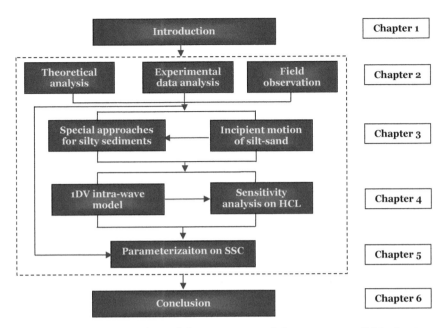

Figure 1-7. *An overview of the contents and the structure of this thesis*

Chapter 2[*]

Sediment transport in silt-dominated coastal areas: Field works and data analysis

Field observations were carried out in the northwestern Caofeidian sea area, a silt-dominate coast, in Bohai Bay, China. The characteristics of tidal currents, waves, and SSC changes under different hydrological conditions were analyzed. Results show that the SSC is mainly influenced by wave-induced sediment suspension: under light wind conditions the SSC was very low, with the peak value generally less than 0.1 kg/m³; the SSC increased continuously under the gales over 6-7 in Beaufort scale, with sustained wind action. The measured peak SSC at 0.4 m above the seabed was 0.15-0.32 kg/m³, with the average value of 0.08-0.18 kg/m³, which is about 3-6 times the value under light wind conditions. Field data in other silt-dominated coasts was collected, such as Huanghua port sea area, Jingtang port sea area and Jiangsu coast. Results show that silt-dominated sediments are sensitive to flow dynamics: the SSCs increase rapidly under strong flow dynamics (i.e., strong tidal currents or waves), and higher concentration exists near the bottom; as a result, the high SSC causes heavy sudden siltation in navigation channels.

---

[*]Parts of this chapter have been published in: **Zuo, L.,** Lu, Y., Wang, Y. and Liu, H., 2014. Field observation and analysis of wave-current-sediment movement in Caofeidian Sea area in the Bohai Bay, China. China Ocean Engineering, 28(3): 331-348.

2.1. Introduction

In coastal and estuarine areas, water and sediment movement are greatly influenced by the interactions of waves and tidal currents. In order to study the flow and sediment movement, it is essential to measure the hydrodynamic and sediment transport processes with high spatial and temporal resolutions. This is valuable not only in practical engineering but also in theoretical research. However, it is difficult to measure the details of flow-sediment process near the bottom by traditional instruments, such as propeller flow velocity meter, which can work only at a single point or multiple points.

In recent decades, significant progress has been made in in-situ flow-sediment measurement. Some of the most popular instruments include Acoustic Doppler Current Profiler (ADCP), three-dimensional, high-frequency Acoustic Doppler Velocimeter (ADV), Optical Backscatter Point Sensor (OBS), Laser In-Situ Scattering and Transmissometry (LISST-100), etc. Sternberg (1968) reported water and sediment movement measured with a tripod system in the tidal channel of Puget Sound. Based on the observed near-bottom velocity distribution, the bottom shear stress and drag coefficient were calculated, and the formulas derived from lab and river experiments was tested in tidal channels. Since then, the United States Geological Survey (USGS), Virginia Institute of Marine Science (VIMS), National Oceanic and Atmospheric Administration (NOAA), among others, have developed near-bottom observation systems (Cacchione et al., 2006) and organized a series of large-scale observation projects. These projects include CODE (Coastal Ocean Dynamics Experiment) (Cacchione et al., 1987; Grant et al., 1984), STRESS (Sediment Transport on Shelves and Slopes) (Sherwood et al., 1994), STRATAFORM (Nittrouer, 1999), and EuroSTRATAFORM (Fain et al., 2007; Nittrouer et al., 2004). Some scholars carried out boundary layer water and sediment observations in different coastal areas and collected waves, currents, and sediment concentration data under normal or stormy weather conditions (Madsen and Wood, 1993; McClennen, 1973; Williams et al., 1999; Wright et al., 1991).

Zhao and Han (2007), Yang and Hou (2004), Sun et al. (2010) and Wang et al. (2012) collected field data in Huanghua port, Jingtang port and Jiangsu coast, and analyzed the sediment transport in silt-dominated coasts. These data were also cited in this section to describe general sediment transport of silt-dominate sediments.

This section presents a field observation carried out in the northwest coastal area of Caofeidian, Bohai Bay, as shown in Figure 2-1. The hydrometric measurement in 2010 was also collected, which was done by Yangtze River Estuary Hydrology and Water Resources Survey Bureau. Tidal levels, velocities and suspended sediment concentration were obtained during one spring tide or neap tide. However, the SSC in 2010 was only measured in light wind conditions; furthermore, there were only six points in one vertical line, which cannot be used for analysing the flow-sediment dynamics near the bed bottom. We need more measured data with high resolution and long-time series to find the changes of suspended sediment concentration under different conditions. Measurement of tidal currents, waves and SSC in the bottom boundary layer during a single tidal period was carried out. Long-term observations were also carried out to evaluate changes of SSC under different hydrologic conditions. Sediment density distribution was detected by a γ-ray densitometer to check whether fluid mud exits in this area.

2.2. Field works and data analysis in northwestern Caofeidian sea area of Bohai bay

2.2.1. Observation site and the tripod system

Human activities are intensive in the Bohai Bay, for example, reclamations of the Caofeidian Port and Tianjin Port. In order to study the natural flow and sediment movement, it is better to choose a location with limited influence of human activities. Therefore, an observation site between the Jianhe River estuary and Heiyanzi Shahe River estuary is selected, which locates in the northwestern sea area of Caofeidian. It is near the two-m isobath (under the lowest tidal level), about nine km away from the shore (Figure 2-1).

A two meters high tripod observation system was installed in the selected site as shown in Figure 2-2. It consisted of AWAC, ADCP, OBS-3A and ADV with different observation objectives, as well as other auxiliary parts such as anchors, ropes, floats, and flashing light. The tripod system had a narrow upper part and a wide lower part, and the three bases were fixed by steel nails anchoring 50 cm into the seabed.

Different instruments were installed on the side and top bars at different heights. The first level, located 0.4 m away from the bottom, was equipped with an ADV for near-bottom, high-frequency velocity measurement and turbulent field analysis, and an OBS with probes (turbidity, salinity, and temperature) mainly for observing bottom sediment concentration. The second level, 1.8 m away from the bottom, included an ADCP overlooking seabed for observing near-bottom instantaneous velocity profile, an AWAC looking up for observing wave elements and instantaneous velocity profile, and another OBS. Accurate measurements of the mean flow characteristics and high-frequency turbulence in the wave-current BBL were carried out using all the instruments installed on the tripod system in one tidal period during October 25-26, 2011. The suspended sediment concentration process was observed for a longer period from October 25 to December 7, 2011. Due to the limitation of the battery capacity of the ADV and ADCP, only the OBS and AWAC were used to observe the sediment concentration and wave characteristics over longer period.

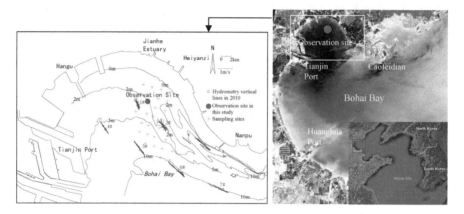

Figure 2-1. *Location of the tripod system observation site and sampling sites*

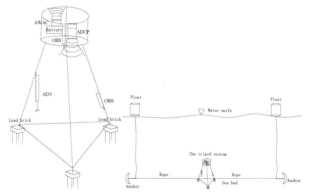

Figure 2-2. *The tripod system and its arrangement*

2.2.2. Bed materials

Caofeidian sea area is a silty-muddy coast (Lu et al., 2009). Its east side, the Jingtao port sea area, is silty coast; while its west side, the Tianjin port sea area, is muddy coast. Based on bed materials sampling in 2006, 2010 and 2011 (Figure 2-3 and Figure 2-4), the seabed sediment distribution is characterized by fine, coarse, and fine moving from land to sea. With the Caofeidian foreland as the dividing line, the grain size increases from west (0.008–0.027 mm or 5–7φ) to east (0.012–0.250 mm or 2–6φ), where φ scale is defined as $d_{50}=2^{-\varphi}$ with d_{50} is the median diameter of the sediment. The median diameter of sediments in the east may be several times larger than that in the west. In the west, the sediment deposits are moderately sorted. In the east, moving from cost to sea, the sediment deposits are moderately, well, and moderately sorted. The median grain size is larger on the shoal than in the deep channel. In the sea area in front of the barrier island, east of the Caofeidian, the sediment deposits are very well sorted, indicating the presence of the strongest waves (Lu et al., 2009).

In the observation sea area, the northwest part of Caofeidian, the bed materials mainly vary from clay silt to coarse silt, with median grain sizes of 0.01–0.05 mm. Close to Nanpu, the bed materials are coarser with most part being very fine sand, because there are tidal sand ridges in this area. Silt is dominating composition in the study area, accounting for 40-70%; generally, the sand composition mostly are less than 20% and the clay composition is 10-20% on average.

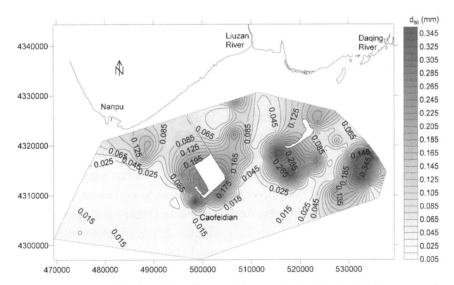

Figure 2-3. *Distribution of medium sediment size in the Caofeidian coastal area measured in 2006*

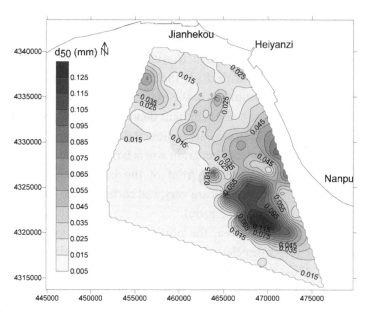

Figure 2-4. Distribution of medium sediment size in the northwestern Caofeidian coastal area measured in 2010 and 2011

2.2.3. Tides, waves and current velocity

Tidal currents

The Caofeidian coastal area is mainly controlled by the Bohai tide system. The tides are irregular semidiurnal with a westward increasing mean tidal range. At the Caofeidian Island, the mean tidal range is 1.54 m (Lu et al., 2009). According to the measured data of June 2010, the mean and maximum tidal range in Jianhe estuary, west of Caofeidian, was 2.45 m and 3.58 m, respectively.

In the study area, the tidal currents in the open sea has reciprocating flow characteristics, with the flow direction of southeast to northwest during flood tide and northwest to southeast during ebb tide (Figure 2-1); however, the tidal currents in near shore area have rotational flow characteristics, within two-m isobath of water depth. The measured mean velocity in June 2010 was 0.39 m/s for the flood currents and 0.32 m/s for the ebb currents during the spring tides, respectively.

Figure 2-5 shows the observed flow depth, velocity, and flow direction during one tidal period. The water depth at the observation site was 2.2–5.3 m. The ebb tide range during the observation period was 2.57–2.19 m with an average of 2.38 m, and the flood tide range were 2.56–2.89 m with an average of 2.72 m. The flood and ebb tide durations were 6 hr 57 min and 5 hr 30 min, respectively. The tidal direction at the observation point had rotating flow characteristics, swirling anticlockwise. The tidal waves were standing waves, with slack water happening around high and low tides and the maximum velocity generally occurring at the half tide. Based on the flow profiles observed by ADCP and AWAC, the maximum velocity was 0.57 m/s during flood tide and 0.36 m/s during ebb tide. The average velocity observed with ADV at 0.4 m above the bottom was less than 0.5 m/s, and the velocity during flood tide was slightly higher than that during ebb tide.

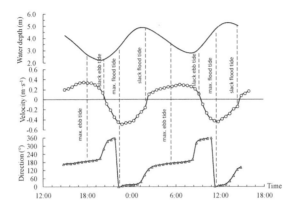

Figure 2-5. *Measured water depth, velocity, and flow direction during one tidal period*

Waves

The significant wave heights observed during this tidal period were 0.12–0.46 m, and the 1/10 significant wave heights were 0.15–0.61 m (Figure 2-6). The main wave directions were SSE and SSW, with average wave direction being 160.9–232.7°. The wave periods were 2–3 s. The angles between waves and currents were 28–44° during flood tide and 124–167° during ebb tide.

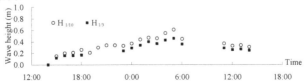

Figure 2-6. *Measured wave height during one tidal period*

Flow structure

The data measured by ADV were used to analyze the characteristics of flow turbulence at the observation site. The ADV was located 0.4 m above the seabed, with a sampling frequency of 8 Hz. The fluctuating velocities in a steady flow are generally obtained by subtracting the average velocity from the instantaneous velocity, as the average velocity in steady flow does not vary over time. Half of the tidal period in the observed sea area lasts for about 6 hours, and the difference between the maximum and minimum velocities can reach 1 m/s. Thus, the average rate of change in tidal velocity is only about 0.0028 m/s in one minute, and the tidal velocity within an interval of one minute can be assumed as approximately constant. In the course of data processing, 480 instantaneous measured velocities within one minute were averaged to arrive at the average velocity and then the fluctuating velocities were obtained by subtracting the average velocity from instantaneous velocities.

The measured data, shown in Figure 2-7, show that the amplitude of fluctuating velocity in N (North) direction is the biggest, followed by that in E (East) direction and then in U (Up) direction. The fluctuating velocities are responsive to the average velocities. When the average velocity is relatively high or low, the fluctuating velocity at the corresponding moment is also high or low. The Reynolds stress and velocity also show a direct correlation that is for high or low velocity, the Reynolds stress is also high or low. The Reynolds stress during flood tide is higher than that during ebb tide. Here, the Reynolds stress is defined as $\tau_{Re} = \rho\sqrt{\left(-\overline{u'w'}\right)^2 + \left(-\overline{v'w'}\right)^2}$, in which u and v are the velocities in the horizontal directions and w is the vertical velocity.

Velocity profile

The vertical velocity distribution was obtained through the combination of bottom velocities measured with ADCP and upper velocities measured with AWAC. It is well known that, as the high frequency vibrating bottom flow induced by waves cannot fully develop, a very thin wave boundary layer is formed at the bottom; on the other hand, the current boundary layer may occupy a large depth range under relatively constant flow. The boundary layers formed by these two dynamics have different characteristics and exert nonlinear influences on each other. Figure 2-8 shows the instantaneous velocity profiles during flood tide, ebb tide, slack flood tide, and slack ebb tide, as well as logarithmic fitting. According to the measured data, the wave-current boundary layer can be divided into three parts: the wave control layer (WCL) near the bottom, the transition layer (TL) in the middle, and the flow (current) control layer (FCL) in the upper part. The velocity profiles in the WCL and FCL follow the logarithmic distribution. The transition layer in the middle part has a more complex profile.

The measured velocity profiles show one velocity overshoot at a certain height from the seabed. The upper limit of this height is the upper height of the WCL, normally 0.3–0.5 m, which is much larger than the measured value in laboratory. This is mainly due to the larger roughness and longer wave period in field situations. The velocity in the upper layer is controlled mainly by tidal currents and follows the logarithmic distribution, which accounts for about 90% of the height.

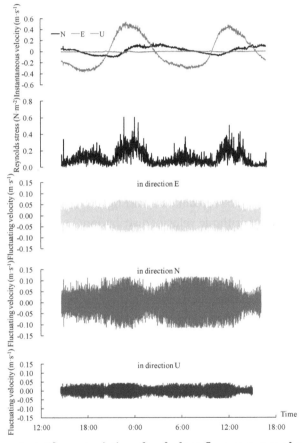

Figure 2-7. *Characteristics of turbulent flow at 0.4 m above the bed*

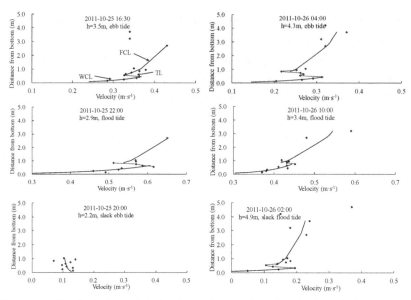

Figure 2-8. *Velocity distribution along the depth (symbols represent measured data and lines represent the logarithmic fitting)*

Bottom shear stress

Bottom shear stress is of great importance to sediment movement. The friction velocity (u_*) can be calculated by fitting the data with the logarithmic velocity profile. The logarithmic velocity distribution $u(z) = u_* / \kappa \ln(z / z_0)$ could be rewritten as $u(z) = u_* / \kappa \ln z - u_* / \kappa \ln z_0 = a \ln z + b$. Then $u_* = \kappa a$ and $z_0 = \exp(b / a)$. In which $u(z)$ is the velocity at height z, κ is Karman number, z_0 is the height where velocity is zero, and u_* is shear velocity. By plotting $u(z)$ and $\ln z$, a and b can be obtained from the slop and intercept of the line, respectively. The shear stress can be obtained by the relation $\tau = \rho u_*^2$.

Figure 2-9 shows the calculated shear stresses from the bottom velocity profile and upper velocity profile. The bottom shear stress during the flood tide was higher than during the ebb tide. The wave shear stress (averaging at 0.71 N/m²) is higher than the current shear stress (averaging at 0.36 N/m²).

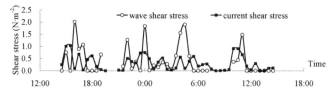

Figure 2-9. *Bottom shear stress due to waves and currents*

2.2.4. Suspended Sediment Transport

The changes of sediment concentration at the observation site, under different meteorological and hydrological conditions from October 25 to December 7, 2011, were measured using OBS.

Figure 2-10 depicts the nearshore wind speed during the observation period. The mean wind speed was about 2.8 m/s. During middle to late November, strong winds occurred with the mean hourly maximum wind speed of 10.8 m/s and extreme wind speed of about 14.8 m/s. The maximum wind forces are 6–7 in Beaufort scale (about 41–62 km/hr) in the late November. The frequent wind directions were W-N and E.

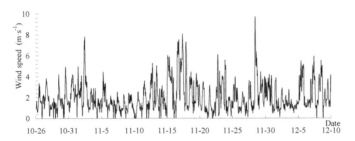

Figure 2-10. *Time series of observed wind speed*

Figure 2-11 shows the time series of significant wave height during the observation period. The AWAC wave data were available until November 15 due to the battery failure. From October 25 to November 15, most of the significant wave heights ($H_{1/3}$) were 0.1–0.5 m, and the $H_{1/10}$ waves were 0.1–0.6 m. The corresponding wave periods $T_{1/3}$ and $T_{1/10}$ were 2–4 s. In order to supplement the wave data, several methods, such as the SMB method (Etemad-Shahidi et al., 2009) and Futaoijima method (Ministry of Transport of China, 1998), are adopted to find out the significant wave heights from the observed wind speed data. Figure 2-11 shows the comparison of the estimated and measured wave heights. Overall, the estimated wave heights match the measured values (except November 9 to 11), with most differences less than 0.1–0.2 m. The Futaoijima method performed better and was chosen to recover the wave data.

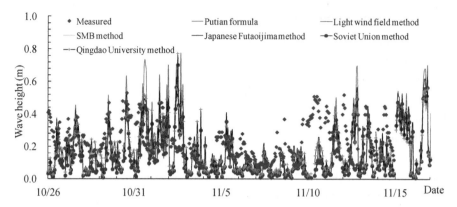

Figure 2-11. *Comparison of estimated and measured wave heights*

Figure 2-12 shows the variation of observed sediment concentration during one tidal period, corresponding to the hydrodynamic process shown in Figures 2-5 and 2-6. The sediment concentrations at 0.4 m above the bottom were 0.01–0.07 kg/m³ with an average value of 0.017 kg/m³. At 1.8 m above the bottom, the sediment concentrations reduced to 0.01–0.04 kg/m³ with an average of 0.013 kg/m³. The sediment concentration is directly correlated to the flow velocity. Based on the formulas of Tang (1963) and Dou and Dong (1995), the sediment incipient velocity near the observation points exceeds 0.5 m/s, but the observed velocities were less than 0.5 m/s in general. Therefore, the sediment concentration is usually small under normal tides and light wind-wave conditions. The sediment concentration during flood tide is higher than that during ebb tide, indicating that sediment mainly comes from the out sea in normal weather days.

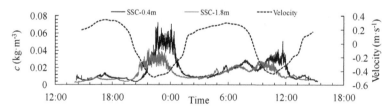

Figure 2-12. *Time series of observed sediment concentrations during one tidal period*

Figure 2-13 shows the time series of SSC measured at 0.4 m above the bottom along with the wave heights. During October 26 to November 13, 2011, the wind speed was low, mostly less than 5 m/s, the wave heights were less than 0.5 m, and the sediment concentration was lower than 0.10 kg/m³ with an average value of 0.03 kg/m³.

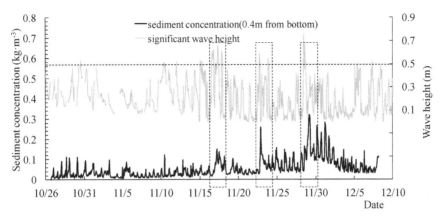

Figure 2-13. Measured suspended sediment concentration along with wave height

In the second half of November 2011, the sediment concentration increased under the action of several strong winds and resulting waves. For example, during November 16–17, when the maximum significant wave heights were 0.55–0.65 m, the measured peak SSC was 0.15 kg/m³ with the average value of 0.08 kg/m³.

During November 22–23 when the maximum significant wave height was about 0.6 m, the peak SSC was 0.26 kg/m³ with the average value of 0.09 kg/m³.

On November 28, when the maximum significant wave heights reached about 0.60–0.75 m, the peak SSC was 0.32 kg/m³ and then decreased to below 0.1 kg/m³. The average sediment concentration during this period was 0.18 kg/m³. However, the peaks of concentration mismatch the peaks of wave heights on November 28-29. The factor may be that the wave height was derived from wind speed by empirical methods, which is likely to have caused the mismatch in phasing between waves (estimated) and concentrations (measured).

It can be concluded that, changes of sediment concentration are basically related to wind waves. The critical wave height for sediment resuspension is about 0.5 m. Under light wind, the peak SSC in the observed sea area was generally less than 0.1 kg/m³, and the average SSC was only 0.03 kg/m³. In strong wind conditions, sediment is stirred up, and the SSC kept increasing under the continuous wind force over 6–7 in Beaufort scale (about 41–62 km/hr). The measured peak SSCs were 0.15–0.32 kg/m³ and the average SSCs during wind-wave action were 0.08–0.18 kg/m³, which were about 3–6 times of the values under light wind conditions.

2.2.5. Water-sediment mixture density

The density of water-sediment mixture near the bed bottom was measured using a γ-ray densitometer from Nanjing Hydraulic Research Institute (Figure 2-14). The density was used to investigate whether fluid mud exists in the study area. Fluid mud is often defined as the water-sediment mixture with density between 1.05 g/cm3 - 1.25 g/cm3, while the density between 1.25 g/cm3 - 1.40 g/cm3 is often called soft mud.

The principle of the γ-ray densitometer

When the γ-ray crosses water-sediment mixture, the intensity will decrease and the attenuation of intensity has an exponent relation with density. The density is obtained by the following formula:

$$\rho_m = \rho + \frac{\rho_s - \rho}{(\mu_s \rho_s - \mu_w \rho) d_\gamma} \ln \frac{CR_w}{CR} \tag{2-1}$$

in which ρ_m = water-sediment mixture density, ρ = water density, ρ_s = sediment density, μ_w = γ-ray absorption coefficient of water, μ_s = γ-ray absorption coefficient of sediment, d_γ = distance from the γ-ray radiation source to the detector probe, CR_w = the value from the detector probe in clear water, and CR = the value from the detector probe in water-sediment mixture or sediment bed.

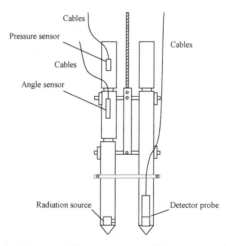

Figure 2-14. *Sketch of the structure of the γ-ray densitometer*

For curtain conditions, as ρ, ρ_s, μ_w, μ_s, d_γ and CR_w are constant, Eq. (2-1) can be simplified as

$$\rho_m = A' + B' \ln \frac{CR_w}{CR} \tag{2-2}$$

in which A' and B' are coefficient. Then the density can be yielded from the detector probe value.

Results

On December 13-14, 2011, density measurement was carried out in the northwestern Caofeidian sea area (Figure 2-15). One day before the measurement, winds with Beaufort scale 6–7 occurred. From the rigid bed to upper part of water column, the density was measured point by point in the vertical line. During measurement, the equipment was slowly input into the water, and the value was recorded at different water depth. When reaching the rigid sea bed, the equipment will incline if it continually goes down, and the angle sensor will send alarm signals when the incline degree is larger than 45°. Then, the measurement will be finished.

The thickness of the fluid mud layer and soft mud layer was analyzed from density distribution. Figure 2-16 shows the distribution of density in the study area. It can be seen that the density at the bed bottom is about 1.4-1.6 g/cm³, and the density decreased to 1.02-1.03 g/cm³ at 0.1-0.2 m above the bed. Table 2-1 and Figure 2-17 shows the thickness of fluid mud and soft mud. It could be seen that the thickness of fluid mud was very small, only 0.0-0.10 m. The thickness of soft mud was only 0.0-0.15 m. There was rarely little fluid mud in the study area.

Table 2-1. Thickness of fluid mud and soft mud in the study area

Sites	Thickness of fluid mud (m)	Thickness of soft mud (m)	Sites	Thickness of fluid mud (m)	Thickness of soft mud (m)
1#	0.10	0.05	21#	0	0
2#	0.05	0.05	24#	0	0.10
3#	0.05	0.10	25#	0.05	0.05
7#	0.10	0.10	26#	0.05	0.10
8#	0.10	0.10	27#	0	0
9#	0.05	0.15	28#	0.05	0.10
10#	0.05	0.05	29#	0.05	0.10
11#	0	0.10	30#	0.05	0.15
14#	0	0	31#	0.05	0.10
15#	0	0	35#	0.10	0.10
16#	0.10	0.05	36#	0.05	0.05
17#	0.05	0.10	37#	0.05	0.15
18#	0	0.10	38#	0	0
19#	0	0	39#	0.05	0.05
20#	0.10	0.05	40#	0.05	0.05

Figure 2-15. The γ -ray densitometer and measurement process

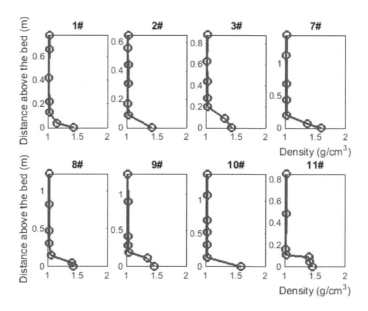

Figure 2-16. Vertical distribution of the water-sediment mixture density in some sample points

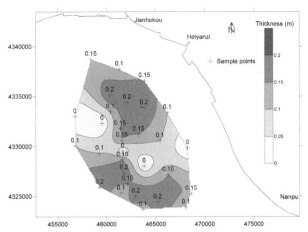

Figure 2-17. *Distribution of the total thickness of fluid mud and soft mud*

2.3. Collected field data in some silt-dominated coasts

Some field data in silt-dominate coasts, such as Huanghua port sea area, Jingtang port sea area and Jiangsu coast, were collected to summarize the sediment movement of silt-dominate sediments.

Huanghua port sea area

Huanghua port locates at southwest of Bohai Bay, see Figure 2-1 and Figure 2-18. The medium size of bed material is 0.005-0.05 mm with the average value of 0.036 mm. The mean tidal range is 2.3 m and the mean water level is 2.4 m. The mean tidal current velocity is about 0.25-0.42 m/s, with the maximum value of about 0.31-0.79 m/s. Measurements show that high SSC occurs during storm surges, which causes heavy sudden siltation in navigation channels.

Yang and Hou (2004) listed SSC with different scenarios of wind forces, see Table 2-2. Figure 2-19 shows the time series of wind speed, significant wave height, wave period and SSC, which were measured in November 5-7, 2001 (Zhao and Han, 2007). All the measured data clearly show that the SSC changes closely relate to wave dynamics/wind forces. Wind-induced waves are main flow dynamics in sediment transport in Huanghua port. The vertical distribution of SSC (Figure 2-20), which was measured during the windy days with the Beaufort scale of about 6, indicates the high sediment concentration near the bottom.

The heavy SSC caused sudden siltation in the Huanghua navigation channel. On October 11-13, 2003, after NE-ENE winds with maximum wind speed of 23.6 m/s, the thickness of back siltation was measured in the outer navigation channel, as shown in Figure 2-21. It can be seen that, the maximum thickness of siltation was 3.5 m, which greatly decreased water depth in the waterway (Zhao and Han, 2007).

Table 2-2. *The SSC under different wind forces in Huanghua navigation channel measured in November 2001 (Yang and Hou, 2004)*

Distance from the coast line	After wind scale of 6	After wind scale of 5	Wind scale less than 5
<5 km	1.20-1.10 kg/m³	0.55-0.45 kg/m³	0.32-0.24 kg/m³
5-12 km	0.81-0.57 kg/m³	0.45-0.25 kg/m³	0.32-0.24 kg/m³

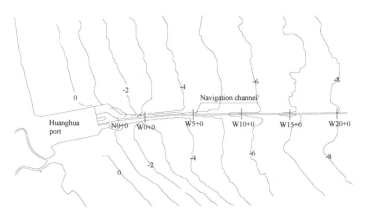

Figure 2-18. *Sketch of the Huanghua port in 2001-2003 (Hou et al., 2013)*

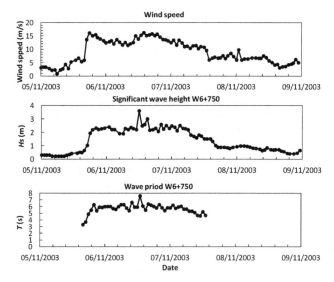

Figure 2-19 (a). *Time series of wind speed, wave height and wave period in November 5-9, 2003* (Zhao and Han, 2007)

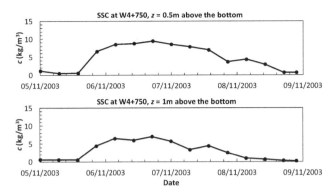

Figure 2-19 (b). *Time series of SSC measured in November 5-9, 2003 (Zhao and Han, 2007)*

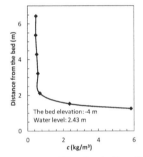 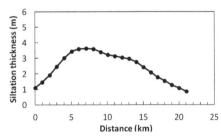

Figure 2-20. *Vertical distribution of SSC during a wind of Beaufort scale 6 measured in March 21, 2003 (Yang and Hou, 2004)*

Figure 2-21. *The thickness of siltation along the Huanghua navigation channel on October 11-13, 2003 (Zhao and Han, 2007)*

Jingtang port sea area (Sun et al., 2010)

The Jingtang port sea area, which locates at east side of Caofeidian in Bohai Bay, is a typical silty coast (Figure 2-22). Near the shore, the medium grain sizes are 0.1-0.2 mm and 0.06-0.09 mm in the area with water depth less than 5 m and 5-8 m, respectively; outside the 8 m in water depth, there are mainly clay silt. The tidal current is relatively weak, with the mean depth averaged current velocity only being 0.2-0.3 m/s. The SSC is similar with Caofeidian sea area and Huanghua port sea area: the SSC is very low in calm wind conditions and increased abruptly in windy days. On October 10-13, 2003, N-ENE winds with the Beaufort scale larger than 6 lasted 34 hours, and the sudden siltation in Jingtang port was heavy, with maximum thickness of 5.5 m and the volume of 1.86 million m³ (Sun et al., 2010). However, during March 2000 to September 2001 under normal conditions, the average thickness of total siltation was only 0.54 m, the maximum thickness was 0.90m, and the siltation volume was 0.095 million m³ in one and half year, which was much less than the sudden siltation in few days.

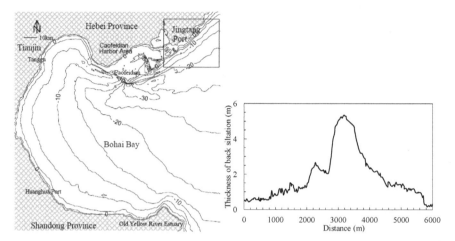

Figure 2-22. *Location of Jingtang port sea area*

Figure 2-23. *Thickness of back siltation along the Jingtang navigation channel (Sun et al., 2010)*

Jiangsu coast (Wang et al., 2012)

The surficial sediment in Jiangsu coast, which locates at north side of Yangtze Estuary, is composed mainly of silts and sandy silts (Figure 2-24), with the medium grain size of 50-82 µm. Field measurement shows that (Wang et al., 2012), the SSCs were characterized by several peaks. Because of the stronger tidal current in Jiangsu coast, these peaks were associated with strong currents, combined wave-current interactions and intense turbulences due to initial flood surge.

Taking M08 as an example (Figure 2-25), which is located at the mid intertidal flat, for the most part the high SSC values (i.e., 0.8-1.6 kg/m³) are associated with high current velocities at the mid-flood and mid-ebb, indicating the occurrence of significant resuspension. The SSC then decreases due to the settling in slack water before the reversal of the tidal current. A peak in SSC (more than 1.5 kg/m³) occurred in strong waves, with significant height of 0.43 m, during the tidal cycles on May 11.

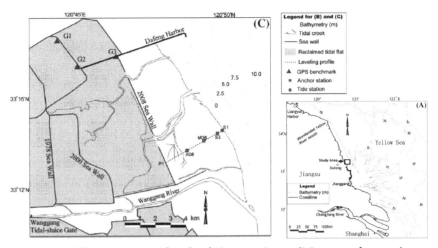

Figure 2-24. *Sketch of Jiangsu Coast (Wang et al., 2012)*

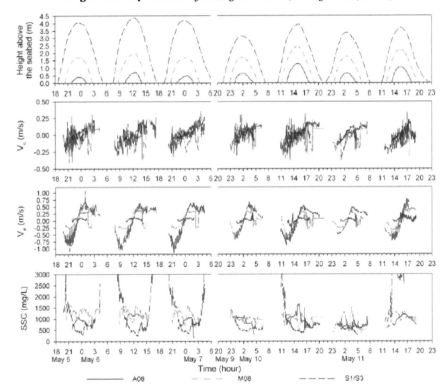

Figure 2-25. *Time series of depth-averaged current velocity and SSC measured at Stations A08 (blue solid-line), M08 (green dash-dot-line), S1/S3 (red dash-line). V_a and V_c are the current velocity components along and across the intertidal flat, respectively (Wang et al., 2012).*

2.4. Summary

(1) Field observations were carried out in the northwestern Caofeidian sea area in the Bohai Bay. Near two-m isobath (under the lowest tidal level), a tripod observation system was installed with AWAC, ADCP, OBS-3A, ADV, etc. Sediment density distribution was detected by a γ-ray densitometer.

(2) Results show that the suspended load sediment concentration is very small under normal tidal and light wind-wave conditions. The sediment concentration increases continuously under the wind force over 6-7 in Beaufort scale, with a sustained wind action. The measured peak sediment concentration at 0.4 m above the sea bed was 0.15-0.32 kg/m³, and the average sediment concentration during wind-wave action was 0.08-0.18 kg/m³, which is about 3-6 times of the value under light-wind conditions. The critical wave height for sediment resuspension is about 0.5 m. The density at the bed bottom was about 1.4-1.6 g/cm³, and decreased to 1.02-1.03 g/cm³ at 0.1-0.2 m above the bed; there was rarely little fluid mud in the study area.

(3) Combined with other field data collected in silt-dominated coasts, such as Huanghua port sea area, Jingtang port sea area and Jiangsu coast, it can be concluded that silt-dominated sediments are sensitive to flow dynamics. Under light flow dynamics the SSCs are normally small; however, the SSCs increase rapidly under strong flow dynamics (i.e., waves or strong tidal currents which can stir up sediments), and show a high concentration layer near the bottom. Meanwhile, because it is easy to settle down, the high concentration caused heavy sudden back siltation in navigation channels.

Chapter 3[*]

Incipient motion of silt-sand under combined action of waves and currents

Sediment incipient motion is a fundamental issue in sediment transport theory and engineering practice. Silt has a transitional behaviour between cohesive and non-cohesive sediment and its incipient motion is still poorly understood. This study aims to find an expression for incipient motion from silt to sand from a unified perspective and analysis. From the analysis of forces, using the derivation method for the Shields curve, an expression for sediment incipient motion is proposed for both silt and sand under conditions of combined waves and currents. The differences and similarities in the sediment motion threshold were analyzed under the effects of waves and currents, as well as fine and coarse sediment. The Shields number was revised by introducing the cohesive force and additional static water pressure, which indicates that this study could be seen as an extension of the Shields curve method for silt. A number of experimental datasets as well as field data were used to verify the formula. The effect of bulk density on fine sediment was discussed and tested using experimental data.

---

[*]This chapter has been published: **Liqin Zuo**, Dano Roelvink, Yongjun Lu, and Shouqian Li., 2017. On incipient motion of silt-sand under combined action of waves and currents. Applied Ocean Research, 69: 116-125.

3.1. Introduction

Sediment movement occurs when instantaneous fluid forces (entraining forces) on a particle are just larger than the instantaneous resisting forces (stabilizing forces) (van Rijn, 1993). This phenomenon is called sediment incipient motion or threshold of sediment motion. Sediment incipient motion is a traditional topic, and is one of the fundamental issues in sediment transport theory and practice.

Sediment grain size is an important factor affecting sediment incipient motion. The initiation of motion of non-cohesive sediments (sand and gravel) has been well studied with both experimental and theoretical works. In contrast, relatively little experimental or theoretical work has been done on the initiation of motion of sediments consisting of cohesive particles (Lick et al., 2004). In particular, the behaviour of silty sediment is poorly understood (Mehta and Lee, 1994). Incipience of motion of cohesive sediment has been studied by some scholars (e.g., Debnath et al., 2007; Kothyari and Jain, 2008; Lick et al., 2004; Righetti and Lucarelli, 2007; Roberts et al., 1998). For cohesive sediments, the cohesive force is much larger than gravity and plays an important role in the resisting forces. Flocculation and consolidation are important physical processes and the floc size and bulk density play a dominant role in controlling the incipient motion conditions of cohesive sediments (Debnath et al., 2007; Lick et al., 2004). Migniot (1968) suggested that the threshold of clay motion follows a direct relation with bulk density.

Since the Shields' curve is not very accurate for fine sediment beds, van Rijn (2007b) proposed empirical calibration factors (cohesive effects and packing effects) for fine sediments. Considering the cohesive force, Tang (1963) and Dou (2000) proposed a critical velocity for fine sediments. Lick et al. (2004) proposed a theoretical description for the fine sediment initiation of motion, including the cohesive forces between particles, as well as changes in bulk density. Righetti and Lucarelli (2007) proposed a threshold criterion for incipient motion of cohesive-adhesive sediments, which was also an extension of the Shields curve.

Mehta and Lee (1994) studied the sediment motion threshold of cohesive materials with $d < 2$ μm and proposed some parameters to describe the incipient motion, such as floc density, solid volume fraction, and cohesive force. However, these parameters do not change gradually within the silt size range; instead, they vary rapidly over a comparatively narrow range, which may possibly be represented by a single size for practical purposes. With respect to the threshold condition for motion, Mehta and Lee (1994) suggested that the

10-20 µm size may be considered to practically be the dividing size differentiating cohesive and cohesionless sediment behavior, while Stevens (1991) proposed that 16 µm was the boundary between sediments that flocculate significantly and those that do not. These theories indicate that the threshold of silt sediment with grain sizes larger than 10-20 µm could be described by grain size, which brings the possibility to extend the cohesionless sediment incipient motion theory to silt sediments.

Earlier studies of the threshold for sediment motion started from uniform flow conditions, and many formulas were proposed. There are two types of formulas: one is based on critical velocity, where the critical condition is expressed by depth-averaged velocity (e.g., Dou, 2000; Maynord, 1978; Tang, 1963), and the other type is based on critical shear stress, where the critical condition is expressed in terms of shear stress; the most widely used is the Shields curve (Shields, 1936; Yalin and Karahan, 1979). Shields (1936) proposed a critical value for the Shields number, $\theta_c = \tau_c / [(\rho_s - \rho)gd]$, as a function of the grain Reynolds number. This theory greatly improved the level of understanding of sediment movement. Here τ_c = critical shear stress, ρ_s = the sediment particle density, ρ = the density of the fluid, g = the acceleration of gravity and d = the sediment particle diameter. The Shields criterion is an empirical relation which is quite general as it applies for any type of fluid, flow, and sediment, as long as the sediment is cohesionless.

Under wave or oscillatory flow conditions, the study of incipient motion of sediment has largely followed the study methods for uniform flow. There are some empirical formulas that have established critical conditions with wave peak orbital velocity (Bagnold and Taylor, 1946; Komar and Miller, 1974; Manohar, 1955; You, 1998). Some scholars, e.g., Dou (2000) and Eagleson et al. (1957) established formulas by combining flume experimental data with theoretical analysis. The Shields curve was extended to wave conditions by some scholars (Madsen and Grant, 1976; Soulsby and Whitehouse, 1997; Tanaka and Van To, 1995).

Under combined wave-current conditions, the formulas for uniform flow or waves are usually applied to determine the critical conditions of sediment incipient motion, however, the dynamic conditions in the formulas need to be changed to those under the combination of waves and currents. There are mainly two categories. The first one is the Shields curve. Experiments and theoretical studies proved that the Shields curve could also be used in wave and wave-current conditions (Li, 2014; Madsen and Grant, 1976; Tanaka and Van To, 1995; Willis, 1978; Zhou et al., 2001). According to van Rijn (1993), initiation of motion in combined current and wave motion can also be expressed in terms of the Shields parameters, provided that the "wave period-averaged absolute bed-shear stress" is used. For the second but more fundamental method, the basic shape of the formula was deduced from a mechanical analysis, and then the coefficients in the formula were determined by experimental data. Bagnold and Taylor (1946), Manohar (1955) and Dou et al. (2001) presented expressions for the critical velocity, critical shear stress, or critical wave height using this method.

From the foregoing review it can be concluded that, although the sediment threshold criterion for motion has been studied extensively and many formulas have been proposed, existing formulas are either limited to flow or wave conditions, or limited to a narrow range of sediment grain size, or too empirical. Therefore, research on sediment incipient motion is still drawing worldwide attention. In natural coastal conditions, waves and currents always coexist. This study aims to find an expression for sediment incipient motion from silt to sand under combined wave-current conditions, which would be generally applicable in modelling and engineering practice.

3.2. Analysis of sediment incipient motion under waves and currents

3.2.1. Similarity and difference of sediment incipient motion between waves and currents

Before deriving the silt-sand incipient motion under combined action of waves and currents, it is necessary to analyze the threshold behaviour of sediments with different grain sizes under different flow dynamics. The differences and similarities in the threshold of sediment motion are analyzed among the effects of waves and currents, and fine and coarse sediment in the following.

(1) Similarities

It is generally accepted that the entraining forces on a sediment grain could be adequately represented by the maximum shear stress generated by a flow, whether this flow is steady (current) or unsteady (wave with or without current). The concept of the threshold for particle motion is also widely accepted for both cohesive and non-cohesive sediments (Righetti and Lucarelli, 2007).

(2) Differences

i) Currents and waves have a different boundary layer and turbulence structure. The critical velocity can only be derived near the bed using mechanical analysis, and then the results of the mechanical analysis is transferred to shear stress or depth-averaged velocity, which are related to the flow structure near the bed. The best unified approach is to use the threshold bed shear stress, which is applicable for conditions with combined or separated waves and currents.

ii) Sediment incipient motion occurs mostly in the rough turbulent region in uniform flow, while under wave conditions, it could happen in the laminar - turbulent region. Tanaka and Van To (1995) collected some experimental data for oscillatory conditions and the results showed that most of the data fall within the laminar-transitional-rough turbulent region.

iii) The forces acting on particles are different for uniform flow and waves. Beside the normal surface drag force and lift force, there is an extra inertia-force of waves in oscillatory flow (Nielsen, 1992).

iv) The stabilizing forces are different for silt and sand particles. For cohesionless sediments the main resistance to erosion is provided by the submerged weight of sediment. However, in a cohesive sediment bed, the resistance is controlled by the net attractive inter-particle surface forces and frictional interlocking of grain aggregates (Kothyari and Jain, 2008). Cohesive forces are important when the bed consists of appreciable amounts of silt particles (van Rijn, 1993). For silty sediments, the correct non-dimensionalization would involve a balance between cohesive forces, shear forces, and gravitational forces (Mehta and Lee, 1994). As the consolidation process is slower when the grain size is smaller, the bulk density would be more important for silt than it is for coarse sand. If these factors are not considered thoroughly in developing the formula, its application will be limited.

Through a thorough analysis of both fluid dynamics and mechanical forces, a unified formula may be obtained for coarse and fine sediments as well as wave-current combinations. In doing so, maintaining simplicity is very important, and, hence, the main factors must be considered and trivial terms must be ignored.

3.2.2. Mechanics analysis of sediment particles

The mechanics of sediment motion have been studied by many scholars (e.g., Dou, 2000; Righetti and Lucarelli, 2007; Shields, 1936; Tang, 1963; van Rijn, 1993). Coarse sediment usually starts to move as single particles while fine sediment tends to move as a group. The various forces acting on particle groups are increased compared with those on single particles, but the balance of the forces or moments on the particle groups can still be treated as that for single particles (Dou, 2000). Besides, the sediment particles are assumed to be ideal ellipsoids, although this is not generally the case. The offset could be corrected by empirical coefficients from experimental data. Under current or wave conditions, the driving forces on sediment motion include the drag force, lift force, and wave inertia-force; the stabilizing forces keeping sediments stable include the submerged gravity, cohesive force between particles, and additional static water pressure (Figure 3-1).

1) Drag force, F_D

$$F_D = C_D a_1 \frac{\rho}{2} d^2 u_0^2 \tag{3-1}$$

The drag force includes the shape drag force, F_{Df}, and the surface drag force, F_{Ds}, where u_0 is the flow velocity near the sediment particle, C_D is the drag force coefficient, ρ is the water density, and a_1 is coefficient.

2) Lift force, F_L

$$F_L = C_L a_2 \frac{\rho}{2} d^2 u_0^2 \tag{3-2}$$

The lift force is mainly caused by flow asymmetry. Where C_L is the lift force coefficient, and a_2 is coefficient.

3) Gravity (submerged particle weight)

$$W = a_3 (\rho_s - \rho) g d^3 \tag{3-3}$$

in which a_3 is the coefficient of sediment particle volume.

4) Cohesive force, F_c, and additional static water pressure, F_δ

An attraction force, F_c, and additional static pressure, F_δ, exist among particles (Dou, 2000; Tang, 1963). The cohesive force is related to the particle size, and physical and chemical properties of the sediment. It has been shown theoretically and experimentally that the cohesive force between two spherical particles is proportional to the particle diameter, i.e., $F_c \propto d$ (Israelachvili, 1992). Actually, cohesive forces exist in all sediment particles. For coarse sediment, the cohesive force is far smaller than gravity, which is why it is often called non-cohesive sediment; while for very fine sediment, the cohesive force is far larger than gravity, and, therefore, the sediment is called cohesive sediment. The cohesive force is the main force when $d \leq 0.062$ mm (van Rijn, 1993). In addition, there is bound water on the sediment particles' surface, which does not transfer the hydrostatic pressure and has a kind of solid-body characteristic. The pressure transfer does not comply with the law of isotropy, so the contact area of two particles receives additional static water pressure.

Dou (1962) first confirmed the additional static water pressure in a two-quartz-wire test. Wan and Song (1990) conducted a pipe test which showed a considerable effect of the pressure on the incipient motion of fine sediments. The cohesive force and additional static water pressure for fully compacted sediment can be expressed as (Dou, 2000):

$$F_c = a_4 \rho d \varepsilon_0 \tag{3-4}$$

$$F_\delta = a_5 \rho g h d \delta_s \sqrt{\delta_s / d} \tag{3-5}$$

where ε_0 is the parameter of cohesive force, which is related to the particle materials, δ_s = 2.31×10⁻⁷ m is the thickness of the bound water (water film), h is the water depth, and a_4 and a_5 are coefficients. Considering moment balance and assuming $a_4 = \alpha' a_5$, $F_c + F_\delta = a_5 (\rho d \varepsilon_k + \rho g h d \delta_s \sqrt{\delta_s / d})$, in which $\varepsilon_k = \alpha' \varepsilon_0$ = 1.75×10⁻⁶ m³/s² based on experimental data for natural sediments (Dou et al., 2001). During the derivation procedure in Sect. 3.3.1, $F_c + F_\delta$ was treated as a whole.

In the silt sediment range, the cohesive force, F_c, is the main force, but the contribution of F_δ becomes bigger as the grain size decreases. The ratio of F_δ / F_c ranges from 15.8% (for d = 0.062 mm, h = 2 m) to 27.8% (for d = 0.002 mm, h = 2 m), which indicates that the additional static water pressure, F_δ, is a non-negligible factor for fine particle sizes.

In summary, the stabilizing force includes the immersed gravity, the cohesive force, and the additional static water pressure. The ratio of the cohesive force and additional static water pressure to the total stabilizing force versus sediment grain size d is shown in Figure 3-2. It can be concluded that the expressions of the cohesive force and additional static water pressure are reasonable and can automatically suggest which are the dominating forces according to sediment grain size. The submerged gravity is the main force when $d > 0.5$ mm; the cohesive force and additional static water pressure are the main forces when $d < 0.03$ mm; and gravity and cohesion are both important when $0.03 < d < 0.5$ mm.

5) Wave inertia-force, F_V (Nielsen, 1992; Zhou et al., 2001)

When calculating the pressure force on an object which is held fixed while the fluid is accelerating flowing past it, an extra mass must be added corresponding to the volume of surrounding fluid which the object keeps from accelerating (Nielsen, 1992). $F_V = F_P + F_H$, where the component F_P is caused by the water column pressure gradient and F_H is caused by the hydrodynamic virtual mass of particles. $F_P = \rho \pi / 6 d^3 du_\infty / dt$ and $F_H = C_M \rho \pi / 6 d^3 du_\infty / dt$. Here, u_∞ is the horizontal free stream velocity and C_M is coefficient.

According to the literature, the ratio of wave inertia force to drag force is proportional to d/A (Nielsen, 1992), where A is the wave amplitude. Experiments and calculations have shown that, this ratio is small, only about 2×10^{-2} in laminar flow and 0.009-0.09 in turbulent flow (Zhou et al., 2001), which means that the wave inertia force can be ignored.

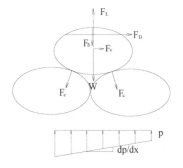

Figure 3-1. *Forces acting on a single particle*

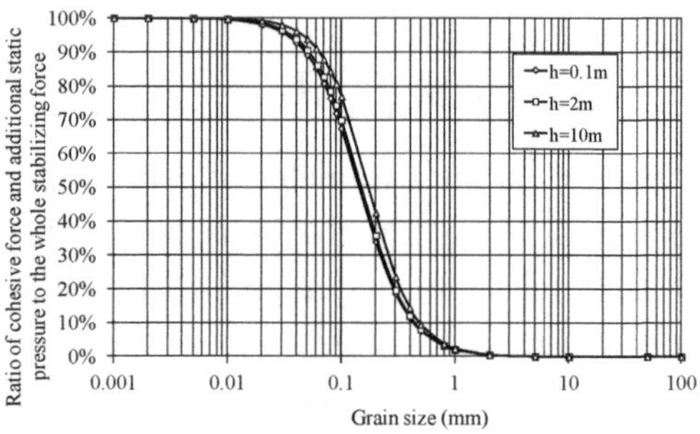

Figure 3-2. *The percentage of cohesive force and additional static water pressure in the total stabilizing force*

Thus, with the wave inertia-force ignored, the driving forces are similar for uniform flow, waves, and combined wave-current conditions. Considering all stabilizing forces, including submerged gravity, cohesive force and additional static water pressure, a unified view of the resisting forces for silt and sand sediments could be achieved.

3.3. Derivation

3.3.1. Theoretical derivation

According to the momentum balance of driving forces and stabilizing forces:

$$K_1 dF_D + K_2 dF_L = K_3 dW + K_4 d(F_c + F_\delta) \tag{3-6}$$

where, K_1 is the coefficient of the movement arm of the drag force, K_2 is the coefficient of the movement arm of the lift force, K_3 is the coefficient of the movement arm of gravity, and K_4 is the coefficient of the movement arm of the cohesive force and additional static water pressure.

Substituting the force expressions of Eqs. (3-1) - (3-5) into Eq. (3-6) yields:

$$u_{0,cr} = \alpha_1 \sqrt{(s-1)gd + a\frac{\varepsilon_k + gh\delta_s\sqrt{\delta_s/d}}{d}} \tag{3-7}$$

in which $\alpha_1 = \sqrt{2K_3 a_3 / (K_1 C_D a_1 + K_2 C_L a_2)}$, $u_{0,cr}$ is the critical velocity on the particles near the bed, $s = (\rho_s - \rho)/\rho$ is the relative density. $a = K_4 a_5 / K_3 a_3$ is a coefficient, which is the ratio of the movement arm of the cohesive force against the movement arm of gravity, and according to the experimental data a = 0.19 (Dou, 2000). It is preferred to use the critical shear stress as the characteristic parameter for incipient motion, as using the near-bed velocity is not practical. Then, u_0 could be further transferred to shear stress based on the flow structure near the bed.

In the turbulent regime, the velocity distributions meet the logarithmic law near the bed in wave, current, or wave-current boundary layers. Here Einstein's formula can be applied to describe the logarithmic layer, which summarizes the flow regimes of smooth, transition, and turbulence.

$$\frac{u}{u_*} = 2.5\ln(30.2\frac{z\chi}{k_s}) = f_1(\mathrm{Re}_*) \tag{3-8}$$

where k_s is the roughness height, u_* is the shear velocity, χ is a correction parameter, $\chi = f(k_s/\delta) = f(\mathrm{Re}_*)$, z is the elevation above the bed, δ is the thickness of the laminar layer near wall, and Re_* is the shear Reynolds number.

In the laminar regime or sub-layer near the bed, the velocity distribution is linear. The velocity distribution is

$$\frac{u}{u_*} = \frac{u_* z}{v} = f_2(\mathrm{Re}_*) \tag{3-9}$$

It can be seen that, the velocity distribution can be represented by shear velocity u_* under different functions of Re_* for laminar-turbulent flow.

From Eq. (3-8), $u_{0,cr} = u_* f_1(\mathrm{Re}_*)$, substitute to Eq. (3-7), then

$$\frac{u_*}{\sqrt{(s-1)gd + a\dfrac{\varepsilon_k + gh\delta_s\sqrt{\delta_s/d}}{d}}} = \frac{\alpha_1}{f_1(\mathrm{Re}_*)}$$

While from Eq. (3-9), $u_{0,cr} = u_* f_2(\mathrm{Re}_*)$, then

$$\frac{u_*}{\sqrt{(s-1)gd + a\dfrac{\varepsilon_k + gh\delta_s\sqrt{\delta_s/d}}{d}}} = \frac{\alpha_1}{f_2(\mathrm{Re}_*)}$$

Combination with $\tau = \rho u_*^2$ and use $f(\mathrm{Re}_*)$ to represent $\dfrac{\alpha_1^2}{f_1^2(\mathrm{Re}_*)}$ and $\dfrac{\alpha_1^2}{f_2^2(\mathrm{Re}_*)}$, yields

$$\frac{\tau_c}{\rho(s-1)gd + a\rho\dfrac{\varepsilon_k + gh\delta_s\sqrt{\delta_s/d}}{d}} = f(\mathrm{Re}_*) \tag{3-10}$$

Rewriting Eq. (3-10), the sediment incipient motion expression is $\theta_{zc} = f(\mathrm{Re}_*)$.

In which,

$$\theta_{zc} = \frac{\tau_c}{\rho(s-1)gd + a\rho \dfrac{\varepsilon_k + gh\delta_s \sqrt{\delta_s / d}}{d}},$$

which represents the ratio of the drag forces to the resisting forces. θ_{zc} is named the incipience number here, which has a similar physical meaning as the Shields number,

$$\theta_c = \frac{\tau_c}{\rho(s-1)gd},$$

and could be seen as an extended version of the Shields number by introducing the cohesive force and additional static water pressure. τ_c is the critical shear stress, which could be the uniform flow shear stress or wave maximum shear stress, or the shear stress of combined waves and currents. $f(\mathrm{Re}_*)$ is the function of $u*$ and is determined by experimental data fitting.

The use of the shear velocity, $u*$, and the bed shear stress, τ_c, on both the abscissa and ordinate can cause difficulties in interpretation, since they are interchangeable (Beheshti and Ataie-Ashtiani, 2008), which has been discussed by many scholars regarding the Shields curve (e.g., Hanson and Camenen, 2007; Madsen and Grant, 1976; Soulsby and Whitehouse, 1997; van Rijn, 1993; Yalin and Karahan, 1979). In this study, following Madsen and Grant (1976), the non-dimensional sediment Reynolds number $\mathrm{Re}d_* = d / 4\upsilon\sqrt{(s-1)gd}$ was used for the abscissa, where υ is the kinematic viscosity of water. Actually, $\mathrm{Re}d*$ was proposed for cohesionless sediments. In this study, it is applied to the silt range. The relation between θ_{zc} and $\mathrm{Re}d*$ was determined by experimental data.

Theoretically, sediment motion initiation is influenced by the flow regime. The friction coefficient in the laminar flow and the smooth turbulent flow regimes is larger than that in rough turbulent flow regime. This is one of the reasons why the sediment critical shear stress is larger when the Reynolds number is lower. In addition, when the sediment size is less than δ (the thickness of the laminar layer near the bed), $\delta = 11.6\upsilon / u*$, the sediment particle is hard to move as it is shielded by the near-bottom flow from the turbulence. Thus, in the smooth turbulent flow regime, the flow around sediment is similar to that with laminar flow. This means that the critical shear stress for fine sediment is higher. These concepts have been accepted widely (Yalin and Karahan, 1979; Zhou et al., 2001). Considering Eq. (3-8), in the smooth turbulent flow regime, χ, C_D, C_L are functions of the Reynolds number, and θ_{zc} varies with $\mathrm{Re}*$. As the χ is in the denominator, θ_{zc} has an inverse relation with $\mathrm{Re}*$. Based on a theoretical analysis, Li et al. (2014) proved that in the laminar or smooth turbulent regimes, the relation between the Shields number and Red_* is $\theta_c = m\,\mathrm{Re}\,d_*^n$, where m and n are coefficients. For the Shields curve, this relation has been investigated by many experiments, but the coefficients are different according to different scholars (Shields, 1936; Yalin and Karahan, 1979; Zhou et al., 2001). For coarse sediment, the flow regime is mostly turbulent, i.e., $\chi = 1$. With constant coefficients of the drag force, C_D, and the lift force, C_L, then θ_{zc} is a constant too; meanwhile the cohesive force can be ignored as it is far smaller than gravity, so Eq. (3-10) will have the same expression as Shields curve for coarse sand. Thus, from above theoretical analysis, θ_{zc} has the following expression as a function of Red_*.

$$\theta_{zc} = \begin{cases} m\,\mathrm{Re}\,d_*^n & \mathrm{Re}\,d_* < d_1 \\ f(\mathrm{Re}\,d_*) & d_1 \leq \mathrm{Re}\,d_* \leq d_2 \\ b & \mathrm{Re}\,d_* > d_2 \end{cases} \tag{3-11}$$

in which m, n, and b are coefficients and $f(\mathrm{Re}\,d_*)$ is an uncertain expression, which can be determined using experimental data. d_1 and d_2 are the inflection points of Red_*.

3.3.2. Experimental data fitting

Laboratory experiments were carried out to determine the fine sediment incipience of motion in a wave flume at Nanjing Hydraulic Research Institute, Nanjing, China. The wave flume is 175 m long, 1.6 m wide, and 1.2 m deep (Figure 3-3). The wave generator was located at one end of the flume, and gravel wave absorbers were positioned to minimize the reflection of the wave. A bidirectional current can be generated by means of pumps. The water surface elevation was measured by Wave Height Meters (WHMs) and the current velocity was measured by an Acoustic Doppler Velocimeter (ADV). The medium grain size was 0.068 mm and 0.125 mm for different sets of experiments. The sediment test section was 10 m long and 0.1 m thick. To support the sediment bed, there were two concrete ramps at the front and at the end of the sediment bed. Based on the experience (Jiang et al., 2001; Xiao et al., 2009), the sediment was immersed in water for one day before the experiment. The flume was slowly filled to the required depth, and then the pump was opened to generate current. Once the current became steady, waves were generated with incrementally increasing wave height until the sediment began to move. There are two primary methods for the identification of sediment threshold (Miller et al., 1977). One is based upon the "small degree" of sediment transport rate (Neill and Yalin, 1969). The net transport rate under wave conditions is usually very low, thus using the sediment transport rate as the threshold is not appropriate. Another type of threshold criterion is the judgment of the sediment motion as "weak movement", "general bed movement" or "scattered particle movement". The incipience was judged as "weak movement" in this experiment, which was also commonly used in this kind of experiments (Dou et al., 2001; Jiang et al., 2001). These experiments were described in detail by Li et al. (2014). The experimental conditions as well as the results are listed in Table 3-1.

Table 3-1. Sediment and hydraulic parameters of the sediment incipient motion experiments

d_{50}(mm)	h(m)	u_c(m/s)	T(s)	H(cm)	d_{50}(mm)	h(m)	u_c(m/s)	T(s)	H(cm)
0.068	0.305	0.0	2.0	3.18	0.125	0.288	0.0	2.0	5.84
0.068	0.485	0.0	2.0	5.40	0.125	0.505	0.0	2.0	6.52
0.068	0.298	0.084	2.0	2.54	0.125	0.492	-0.140	2.0	3.69
0.068	0.298	0.148	2.0	2.44	0.125	0.292	-0.145	2.0	4.77
0.068	0.293	-0.142	2.0	2.61	0.125	0.294	0.147	2.8	3.12
0.068	0.296	0.140	2.8	2.45	0.125	0.280	-0.142	2.8	2.73
0.068	0.496	0.148	2.0	3.85	0.125	0.500	0.141	2.0	5.23
0.068	0.487	0.235	2.0	3.78	0.125	0.495	-0.144	2.0	4.22

Note: d_{50} is median diameter of sediment grain, h is water depth, u_c is current velocity, T is wave period and H is wave height.

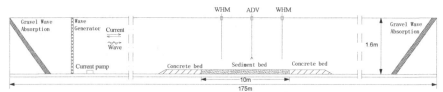

Figure 3-3. Sketch of the experiment flume

Many experimental data were collected for sediment incipient motion, including fine sediment to coarse sediment, current conditions, wave conditions and combined wave-current conditions (Table 3-2). Some experiments which were cited by the Shields curve, Yalin and Karahan (1979) and Soulsby and Whitehouse (1997), were done in glycerol or mentor oil instead of water to achieve a laminar flow regime or by light-weight materials instead of natural sediment. These data were ignored because the cohesive force coefficient in this paper is only available for water and natural sediment. There were also some experimental data carried out in both water and mentor oil for fine sediment by White (1970), which was referred by Soulsby and Whitehouse (1997) and Miller et al. (1977). In White (1970)'s experiments, sediment beds were prepared by depositing the material grain by grain from water moving not quite strongly enough to cause erosion. In the oil experiments, the sediment beds were prepared by raking. It deals only with threshold of sediments that are as near as possible free of the cohesive material. Thus, the data of White (1970) was not used here for calibration.

Table 3-2. *Experimental data for sediment incipient motion*

Source	Hydrodynamic force	Sediment
Smerdon and Beasley (1961)	Current	Silt (d_{50} = 0.005 - 0.02 mm)
Yalin and Karahan (1979)	Current	Sand - gravel (d_{50} = 0.12 - 8.8 mm)
Madsen and Grant (1976)	Wave/ Oscillatory motion	Sand - gravel (d_{50} = 0.09 - 8.0 mm)
Miller et al. (1977)	Current	Sand (d_{50} = 0.09 - 3.5 mm)
Soulsby and Whitehouse (1997)	Current & wave & wave-current	Sand - gravel (d_{50} = 0.062 - 24 mm)
Dou (2000)	Current	Silt (d_{50} = 0.004 - 0.05 mm)
Zhou et al. (2001)	Wave & wave-current	Sand(d_{50} = 0.06 - 0.9 mm)
Cao et al. (2003)	Wave and current	Silt - sand(d_{50} = 0.04 - 0.28 mm)
This paper	Wave & wave-current	Silt - fine sand (d_{50} = 0.068, 0.125 mm)

Notes: For Yalin and Karahan (1979), some data for glycerol are ignored. The data in Madsen and Grant (1976) came from Bagnold and Taylor (1946), Manohar (1955), Horikawa and Watanabe (1967) and Ranee and Warren (1969). The data of Smerdon and Beasley (1961) was obtained from Kothyari and Jain (2008). The data of Miller et al. (1977) was obtained from Grass (1970), Vanoni (1964) and Everts (1973). The data of Soulsby and Whitehouse (1997) was obtained from Katori et al. (1984), Fernandez Luque and Van Beek (1976), Kapdasli and Dyer (1986), Lee-Young and Sleath (1989), Hammond and Collins (1979), Kantardgi (1992), Willis (1978), Ranee and Warren (1969), Rigler and Collins (1983) and Vincent (1957). d_{50} is the median sediment particle diameter.

Some papers did not publish the water depth, and to calculate the additional static pressure, a water depth of 0.3 m was assumed considering that the water depth in most flume experiments was 0.2-0.5 m. The calculation of stabilizing force shows that for fine sediment with diameter of 62 μm under water depth of 0.1 - 0.5 m, the maximum deviation was about 1.31% compared with that under water depth of 0.3 m. The deviation will be bigger with finer sediment, i.e., for the diameter of 20 μm, the maximum deviation was 2.62%. Therefore the assumption of the water depth of 0.3 m has little influence on the results.

The shear stress cannot be measured directly on a mobile bed, so it has to be derived from the formulas. It is reasonable in practice, because when the critical shear stress is calculated the similar method will be employed too. The wave shear stress $\tau_{wm} = 1/2 f_w \rho u_m^2$, in which u_m is the maximum wave orbital velocity, and f_w is the wave friction coefficient. In laminar flow regime, $f_w = 2/\sqrt{\mathrm{Re}_w}$, where $\mathrm{Re}_w = u_* A / v =$ wave Reynolds number and in turbulence flow regime, $f_w = \exp[5.213(d/A)^{0.194} - 5.977]$ according to Jonsson (1966) and Swart (1974). The shear stress under the combined action of wave and current could be calculated using the method of Soulsby (1997). The maximum shear stress of combined wave and current is $\tau_{wc\max} = [(\tau_{wcmean} + \tau_{wm} \cos\phi)^2 + (\tau_{wm} \sin\phi)^2]^{1/2}$, in which $\tau_{wcmean} = \tau_{current}[1 + 1.2(\tau_{wm}/(\tau_{current} + \tau_{wm}))^{3.2}]$ = the mean shear stress of combined wave and current, $\tau_{current}$ = the current shear stress and ϕ = the angle between wave and current.

Figure 3-4 shows the final incipient motion curve as well as the comparison with the experimental data. It can be seen that the experimental data are scattered but still have a rough trend. When realizing the subjectiveness involved in determining the point of sediment incipient motion during experiments, this scatter was considered acceptable. The curve of the incipient motion is in a shape of a band rather than a line. However, for practical purpose, it is preferred to use a line. From Figure 3-4, it can be concluded that when $\mathrm{Re}d_*$ is smaller than 1 (corresponding to $d \approx 0.1$ mm), θ_{zc} has an inverse exponential relation with $\mathrm{Re}d_*$; when $\mathrm{Re}d_*$ is more than 100, θ_{zc} is constant; and in the transition zone, θ_{zc} has a complex relation with $\mathrm{Re}d_*$, and a logarithmic relation is applied here.

Finally, the expression of sediment incipient motion under combined action of waves and currents is:

$$\theta_{zc} = \begin{cases} 0.025 \mathrm{Re}d_*^{-0.07} & \mathrm{Re}d_* < 1 \\ 0.00543 \ln(\mathrm{Re}d_*) + 0.025 & 1 \leq \mathrm{Re}d_* \leq 100 \\ 0.05 & \mathrm{Re}d_* > 100 \end{cases} \tag{3-12}$$

Eq. (3-12) is only for sediment with stable density and the bulk density's effect is discussed in the next section.

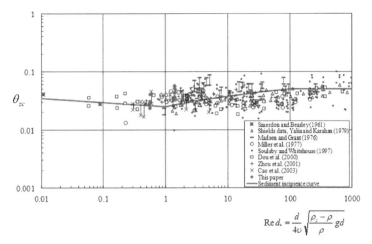

$$\mathrm{Re}\,d_* = \frac{d}{4\upsilon}\sqrt{\frac{\rho_s - \rho}{\rho}gd}$$

Figure 3-4. Sediment incipient motion curve for wave-current conditions

3.4. Verification and discussion

3.4.1. Critical shear stress versus sediment particle size

Figure 3-5 shows the calculated critical shear stress based on Eq. (3-12) as well as a comparison with experimental data and some field data (as listed in Table 3-3). It is very difficult to directly observe the incipient motion of sediment in the field, but an indirect judgment is possible. For sandy seabed, the motion of a sand tracer is often used to show the stage of the incipient motion; for fine sediment sea bed, measurement of the sediment concentration at the sea bottom is usually used for the judgment. From Figure 3-5, it can be concluded that critical stress is strongly dependent on particle size: as the particle size decreases, the critical stress decreases rapidly, reaches a minimum, and then increases. Though there are some deviations, the critical shear stress calculated by Eq. (3-12) shows a reasonable distribution versus sediment size. Furthermore, we have checked the relation of critical shear stress with sediment grain size under water depth of 0.1-30 m and the results show that, the critical shear stress is the smallest when d = 0.1-0.2 mm, at which the W has the same order of magnitude as $F_c + F_\delta$. The particle size corresponding to the minimum critical shear stress is smaller under smaller water depth and vice versa. When the grain size is smaller than 10-20 μm, flocculation may become more important, and the incipient motion is greatly influenced by flocs size and bulk density. Thus, it is worth noting that, though Figure 3-5 shows sediment sizes larger than 2 μm, i.e., the whole silt-sand range, the formulation should be very carefully applied to grain sizes less than 20 μm.

Table 3-3. *Field data for sediment incipient motion*

Source	d (mm)	h (m)	u_c (m/s)	T (s)	H (m)	τ_c (N/m²)
Hill (1963)	0.125	3.0	0	14	0.38	0.43
King (1972)	0.175	2.0	0	3.75	0.25	0.35
Li and Amos (1999)	0.200	56.3	0.143*	9.1	0.73	0.17
Zuo et al. (2014)	0.015	4.0	0.57	3	0.50	0.88

*Note: In Li and Amos (1999), the current velocity is the mean velocity at 50cm above the bed

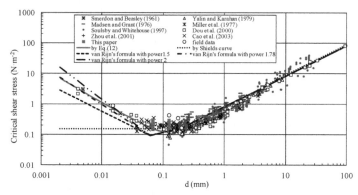

Figure 3-5. *Critical shear stress versus sediment grain size*

3.4.2. Critical shear stress versus bulk density for fine sediments

Some experiments and analyses have shown that (Lick et al., 2004; Roberts et al., 1998), for smaller particles, the critical stresses are strongly dependent on the bulk density with the dependence increasing as the particle size decreases. The bulk density varies greatly for fine sediment depending on the degree of consolidation, which has effects on the cohesive forces and then influences the criterion for incipient motion conditions. Meanwhile, the cohesive force is inversely proportional to particle spacing, meaning that among particles, smaller particle spacing will cause a bigger cohesive force. The sediment bulk density is different with different compactness. Experiments have shown that the smaller the bulk density of fine sediment is, the easier the sediment incipient motion will be (Righetti and Lucarelli, 2007; Roberts et al., 1998), i.e., the critical shear stress increases as the bulk density increases. van Rijn (2007b) assumed that the critical bed-shear stress is affected by cohesive particle-particle interaction effects including the clay coating effects ($\phi_{cohesive}$) and packing (or bulk density) effects ($\phi_{packing}$).

Considering the bulk density, a compaction coefficient, β_c, is introduced and the cohesive force is further expressed as $F_c = \beta_c a_4 \rho d\varepsilon$ and $F_\delta = \beta_c a_5 \rho g h d \delta_s \sqrt{\delta_s / d}$ (Dou et al., 2001). Tang (1963) proposed that the cohesive force is proportional to the ratio of the bulk density to the stable bulk density as a power function of 10. Dou et al. (2001) proposed that

$$\beta_c = \left(\rho_0 / \rho_0*\right)^{2.5},$$

where ρ_0 and $\rho_0* = 0.68\rho_s (d/d_0)^n$ are the dry density and the stable dry density of the sediment respectively, in which $d_0 = 0.001$ m, $n = 0.08 + 0.014(d_{50}/d_{25})$, and d_{25} is the grain size for which 25% of the bed is finer. Details of these parameters can be found in Dou (2000). If the wet density ρ' is used, then $\beta_c = [(\rho' - \rho)/(\rho'* - \rho)]^{2.5}$, in which $\rho'* = \rho + 0.68(\rho_s - \rho)(d/d_0)^n$ is the stable wet density.

Then the compaction coefficient, β_c, is introduced to reflect the compactness degree in Eq. (3-12) using $a \beta_c$ to replace a.

Some experimental data were collected to verify the formula. Figure 3-6 shows the verification of the critical shear stress in Lianyun port (Huang, 1989), La Vilaine estuary and Fodda estuary (Migniot, 1968), Qiantang estuary (Yang and Wang, 1995), and Aojiang estuary (Xiao et al., 2009) under flow or wave conditions. The results show that, the bigger the bulk density was, the higher the critical shear stress was. Figure 3-7 shows the calculated critical shear stress with different β_c. When the sediment size is less than 0.1mm, the shear stress varies a lot (even by several times) with different compaction coefficients. The compaction has nearly no influence on sediments larger than 0.5 mm.

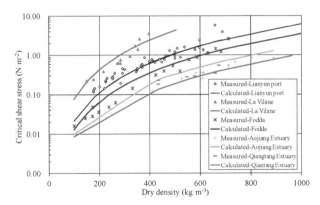

Figure3-6. *Comparison of measured and calculated critical shear stress versus dry bulk density. (a. Lianyun port d_{50} = 0.004 mm, d_{25} = 0.00075 mm, ρ_0* = 767 kg/m³; b. La Vilane estuary, d_{50} = 0.0031 mm, d_{25} = 0.00031 mm, ρ_0* = 506 kg/m³; c. Fodda estuary, d_{50} = 0.0035 mm, d_{25} = 0.0026 mm, ρ_0* = 1030 kg/m³; d. Aojiang estuary, d_{50} = 0.0084 mm, n = 0.14, ρ_0* = 923 kg/m³; e. Qiantangjiang estuary, d_{50} = 0.0104 mm, d_{25} = 0.003 mm, ρ_0* = 1002 kg/m³)*

Although the compaction of fine sediments is very important, its value is hard to determine in the field. It varies in different regions according to the sediment composition and sedimentation history. Further, the bed material distribution in some natural environments is very complex. For example, the presence of organic material may have a great effect on the erosion behavior of fine sediments, and the bed surface may become cemented due to slimes produced by diatoms and bacteria (van Rijn, 2007b). Thus, there should be more studies on the behavior of fine sediment movement and there need more in-situ data on bed materials.

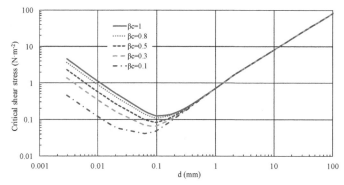

Figure3-7. *Calculated critical shear stress with different compaction coefficients* β_c

3.4.3. Comparison with Shields curve and van Rijn's formula

Figure 3-5 shows the critical shear stress vs. sediment grain size calculated using the Shields curve. The Shields curve was calculated using a fitting expression by Soulsby and Whitehouse (1997):

$$\theta_c = \frac{0.24}{D_*} + 0.055[1 - \exp(-0.020 D_*)] \tag{3-13}$$

in which $D_* = d_{50}[(s-1)g/\upsilon^2]^{1/3}$ = dimensionless particle size.

It can be concluded that, for $d < 0.1$ mm, Eq. (3-13) does not reflect the phenomenon that the critical shear stress becomes higher when the grain size decreases, which has been proven by many experimental data. Considering this problem, van Rijn (2007b) presented the cohesive effects for fine sediments less than 62 μm by

$$\tau_c = \phi_{cohesive} \phi_{packing} \tau_{cr0} \tag{3-14}$$

where $\phi_{cohesive} = (d_{sand}/d_{50})^{\gamma_v}$, $\phi_{packing} = c_{gel}/c_{gel,s}$, with γ_v in the range of 1-2, $d_{sand} = 62$

μm, $c_{gel} = (d_{50}/d_{sand})c_{gel,s}$ = gelling volume concentration of the bed (with $c_{gel,min}$ = 0.05) and $c_{gel,s}$ = 0.65 = maximum volume concentration of a pure sand bed. τ_{cr0} was calculated by a revised Shields curve from the data fitting of White (1970)'s experimental data:

$$\theta_{c0} = \begin{cases} 0.115(D_*)^{-0.5} & D_* < 4 \\ 0.14(D_*)^{-0.64} & 4 \leq D_* < 10 \end{cases} \tag{3-15}$$

in which $\theta_{c0} = \dfrac{\tau_{cr0}}{[(\rho_s - \rho)gd_{50}]}$.

From Figure 3-5, it can be seen that, Eq. (3-12) has a similar tendency and order as van Rijn (2007b)'s formula. It appears that γ_v = 1.78 in van Rijn's formula would yield virtually the similar results as Eq. (3-12), with RMSE = 0.12 (root-mean-square error). If the original Shields curve, Eq. (3-13), was used for τ_{cr0} in van Rijn's formula, γ_v = 1.1 would yield similar results with Eq. (3-12), with RMSE = 0.05.

3.5. Conclusion and remarks

A general expression of sediment incipient motion is proposed for silty to sandy sediments under the combined wave and current conditions. The derivation in this paper follows the route of the Shields curve, and the expression also is similar to Shields curve. Firstly, the differences and similarities of sediment motion initiation are analyzed between wave and current conditions, and between fine and coarse sediments. The differences are considered during derivation, thus, the resulting formulation is valid for both silt and sand sediments and wave-current conditions. The Shields number was revised by adding the cohesive force and additional static pressure, which indicates that this study is an extension of Shields curve. A number of experimental data were used to identify coefficients. Eq. (3-12) shows the final expression for well-compacted sediments. Based on Eq. (3-12), the compaction coefficient is applied in a preliminary way and discussed to reflect the effect of bulk density for fine sediments. Verification and discussion are presented about the dependence of the critical shear stress on sediment particle size and bulk density, which shows satisfactory results.

The finer part of the silt may behave like clay, e.g., eroded in aggregates and the incipience of motion is greatly affected by floc size and the consolidation process. Because flocculation was not considered in this study, the formulation was not suitable for fine sediments with grain sizes less than 10-20 μm, for which the clay theory has to be employed. Chunk or aggregate erosion and the maximum size of the aggregates can be approximately described by equivalent particle sizes. Due to its complexity, the physical-chemical essence of fine sediments still needs further study. In addition, this study only focuses on uniform sediment and the influence of sediment gradation is another important factor in some cases, for example the hiding-exposure effects between different sediment sizes, which has been studied extensively and is a future study direction for simulation of mixed sediment. There are still many other parameters effecting sediment movement, such as mineralogy, organic content, and amounts and sizes of gas bubbles, which may need further study.

Chapter 4

High sediment concentration layer of fine sediments: Modelling by a 1DV model

Experiments and field observations have revealed that when silt and very fine sand are subject to oscillatory wave motion, a high shear flow layer and a high concentration layer (HCL) exist near the bottom. The behaviour of the HCL is still under-researched. Firstly, an intra-wave process based 1DV model was developed for fine sediment transport under the combined action of waves and currents. Some key processes that were included in the model are represented through approaches for different bed forms (rippled bed and 'flat bed'), hindered settling, stratification, reference concentration and critical shear stress. A number of experimental datasets were collected to verify the model, which shows that the model is able to properly simulate the flow and sediment dynamics. Secondly, sensitivity analyses were carried out on some factors which would impact the suspended sediment concentration (SSC) profile of the HCL by the 1DV model. Results show that bed forms play a significant role in the HCL and determination of the shape of the concentration profile. When a current is imposed, the SSC profiles become smoother; however, sediment concentration in the lower HCL is still dominated by the wave motions. For finer sediment, the stratification effects and the mobile bed effects strongly impact the HCL. In conclusion, this paper provides a tool for the study of the HCL and an evaluation of several impact factors on the HCL.

4.1. Introduction

In addition to experiments, bottom boundary layer models (normally 1DV) are powerful tools for studying sediment transport mechanisms. A number of numerical models for sediment transport have been developed over the years. Sediment transport modelling started in the last century with the development of 1DV models (Grant and Madsen, 1979; Grant et al., 1984; Nielsen et al., 1978; Smith and McLean, 1977; Stive and De Vriend, 1994). Because of their simplicity and precision, these models are valuable for some special issues, such as the intra-wave vertical distribution of velocity, shear stress and concentration. Different models have been developed to predict sediment transport under waves or wave-current conditions. These models can be divided into three different classes (Hassan and Ribberink, 2010; Zhang et al., 2011): empirical quasi-steady transport models, intermediate transport models, and fully unsteady sediment transport models (process based). Fully unsteady sediment transport models are based on a time-dependent simulation of both velocities and concentrations during the wave cycle at different elevations above the bed (Fredsøe, 1984; Guizien et al., 2003; Hassan and Ribberink, 2010; Holmedal and Myrhaug, 2006; Kranenburg et al., 2013; Ribberink and Al-Salem, 1995; Ruessink et al., 2009; Uittenbogaard et al., 2001). The process-based unsteady models are based on more advanced approaches, and this study focuses on this kind of model. Up until now, there are many models focusing on sand (e.g., Dong et al., 2013; Kranenburg et al., 2013; Uittenbogaard et al., 2000) and fluid mud (e.g., Hsu et al., 2009; Winterwerp and Uittenbogaard, 1997), but few models on silty sediments.

There is still a lack of thorough modelling and parameterization of the sediment concentration distribution in the HCL of silt and very fine sand. A 1DV model was developed which focuses on silt movement and is also applicable to sand. Process descriptions are given in the model, e.g., different bed forms (rippled bed and 'flat bed'), hindered settling effects, stratification effects, reference concentration and critical shear stress. A number of experimental datasets were collected to validate the model. Sensitivity calculations were carried out by the model to analyze several factors that would impact the sediment concentration profile near the bottom.

4.2. A 1DV model for flow-sediment movement in wave-current BBL

4.2.1. Governing equations

To simulate the intra-wave process of wave-current and sediment concentration, Reynold equations for wave-current BBL were employed.

Following the Reynold's decomposition method, the Reynolds equations for the wave-current boundary layer can be derived from the N-S equations in x-z coordinates. From the derivation process, it is helpful to understand the wave-current interaction terms. Appendix A provides details. The equations read:

Continuity equation

$$\frac{\partial u}{\partial x} + \frac{\partial w}{\partial z} = 0 \tag{4-1}$$

Momentum equation

$$\frac{\partial u}{\partial t} + u\frac{\partial u}{\partial x} + w\frac{\partial u}{\partial z} = -\frac{1}{\rho}\frac{\partial p}{\partial x} + \frac{\partial}{\partial z}[(\upsilon + \upsilon_t)\frac{\partial u}{\partial z}] \tag{4-2}$$

The equation for sediment concentration c

$$\frac{\partial c}{\partial t} + u\frac{\partial c}{\partial x} + (w - w_s)\frac{\partial c}{\partial z} = \frac{\partial}{\partial z}\left(\varepsilon_s \frac{\partial c}{\partial z}\right) \tag{4-3}$$

Here u and w are velocities on x and z coordinates, respectively, p is the water pressure, υ is the kinematic viscosity coefficient, υ_t is the eddy viscosity, c is the sediment concentration, ε_s is the sediment diffusivity, and w_s is the settling velocity.

4.2.2. Turbulence model

The $k - \varepsilon$ turbulence model was employed for eddy viscosity. It consists of transport equations for the turbulent kinetic energy k and the turbulent dissipation ε. For low Reynolds number flow, the NTM model (Sana et al., 2007) with a damping function of standard $k - \varepsilon$ model was employed, which implies that the model is applicable over the entire cross-stream dimension including the low Reynolds number region (viscous sublayer).

$$\frac{\partial k}{\partial t} + u\frac{\partial k}{\partial x} + w\frac{\partial k}{\partial z} = \frac{\partial}{\partial z}\left\{\left(\upsilon + \frac{\upsilon_t}{\sigma_k}\right)\frac{\partial k}{\partial z}\right\} + \upsilon_t\left(\frac{\partial u}{\partial z}\right)^2 - \varepsilon - B_k \tag{4-4}$$

$$\frac{\partial \varepsilon}{\partial t} + u\frac{\partial \varepsilon}{\partial x} + w\frac{\partial \varepsilon}{\partial z} = \frac{\partial}{\partial z}\left\{\left(\upsilon + \frac{\upsilon_t}{\sigma_\varepsilon}\right)\frac{\partial \varepsilon}{\partial z}\right\} + c_{1\varepsilon}f_1\frac{\varepsilon}{k}\upsilon_t\left(\frac{\partial u}{\partial z}\right)^2 - c_{2\varepsilon}f_2\frac{\varepsilon^2}{k} - \frac{\varepsilon}{k}c_{3\varepsilon}B_k \tag{4-5}$$

Here k is the turbulent kinetic energy, ε is the dissipation rate, $\upsilon_t = c_\mu f_\mu k^2 / \varepsilon$ is the eddy viscosity, c_μ = 0.09 is a coefficient, f_μ is a coefficient as listed in Table 4-1, σ_k and σ_ε are turbulent Prandtl-Schmidt numbers for k and ε, respectively; B_k is the buoyancy flux and $c_{3\varepsilon}$ is coefficient, which are related to stratification effects and which will be discussed in section 4.2.4.3.

The various coefficients in the standard $k - \varepsilon$ model and NTM model are summarized in Table 4-1.

Table 4-1. *Coefficients in the standard $k - \varepsilon$ and NTM turbulence model*

	c_μ	f_μ	$c_{1\varepsilon}$	$c_{2\varepsilon}$	σ_k	σ_ε	f_1	f_2
Standard	0.09	1.0	1.44	1.92	1.0	1.3	1.0	1.0
NTM	0.09	$f_{\mu N}$	1.45	1.90	1.4	1.3	1.0	f_{2N}

Note: $\qquad f_{\mu N} = (1 + 4.1 / R_t^{0.75})(1 - \exp(-z^* / 15.75))^2$,

$f_{2N} = (1 - 0.3\exp(-(R_t / 6.5)^2)) \times (1 - \exp(-z^* / 3.64))^2$, $R_t = k^2 / (\varepsilon \upsilon)$, $z^* = (\upsilon \varepsilon)^{1/4} z / \upsilon$.

4.2.3. Approaches for flow simulation

4.2.3.1. Driving forces

The driving forces are pressure gradients outside the BBL. For current-only,

$$-\frac{1}{\rho}\frac{\partial \overline{p}}{\partial x} = -\frac{\partial}{\partial x} g\zeta = gJ \qquad (4\text{-}6)$$

where $\overline{p}$ is the time-averaged pressure, ζ is the water level and J is the mean water surface slope.

For waves-only, the unsteady horizontal pressure gradient $\tilde{p}$ is determined in advance from a given horizontal free stream velocity u_∞ with zero mean:

$$-\frac{1}{\rho}\frac{\partial \tilde{p}}{\partial x} = \frac{\partial u_\infty}{\partial t} + u_\infty \frac{\partial u_\infty}{\partial x} \qquad (4\text{-}7)$$

Thus, for the wave-current cases, the driving force is described as:

$$-\frac{1}{\rho}\frac{\partial p}{\partial x} = -\frac{1}{\rho}\frac{\partial \overline{p}}{\partial x} - \frac{1}{\rho}\frac{\partial \tilde{p}}{\partial x} = gJ + \frac{\partial u_\infty}{\partial t} + u_\infty \frac{\partial u_\infty}{\partial x} \qquad (4\text{-}8)$$

In the governing equations, the Stokes drift is not included as this model is mainly for the BBL, while the Stokes drift mainly affects the velocity near the surface.

4.2.3.2. Simplification of the advection term (1DV-approach)

In order to simplify the mathematical solution to the equations of momentum, continuity and $k - \varepsilon$ model, the relation

$$\frac{\partial}{\partial x} = -\frac{1}{c_e}\frac{\partial}{\partial t} \qquad (4\text{-}9)$$

was applied, where c_e is the wave celerity, x is the horizontal direction and t is the time. Previous work dealing with steady streaming within the ocean BBLs (Deigaard et al., 1999; Holmedal and Myrhaug, 2009; Hsu and Ou, 1994; Trowbridge and Madsen, 1984) have considered boundary layer models where the horizontal gradient operator in the convective term was approximated by this relation. This approximation reduces the two-dimensional boundary layer equations to spatial one-dimensional equations. This kind of boundary layer approximation can only be used in such conditions where the generation of time-dependent turbulence is confined to a relatively thin layer by the short period of the horizontal oscillation compared with the wavelength. More discussions have been presented by Henderson et al. (2004) and Kranenburg et al., (2012).

Using the continuity equation, the vertical velocity at level z can be expressed as:

$$w(z) = \frac{1}{c_e} \int_{0}^{z} \frac{\partial \tilde{u}}{\partial t} dz \qquad (4\text{-}10)$$

where $\tilde{u}$ is oscillatory component of velocity and $w = 0$ at $z = z_0$ is utilized.

4.2.4. Approaches for sediment simulation

4.2.4.1. Settling velocity and hindered settling effects

van Rijn's formula (van Rijn, 1993) was employed for the settling velocity $w_{s,0}$ in clear water,

$$w_{s,0} = \begin{cases} \dfrac{(s-1)gd_s^2}{18\upsilon} & 1 < d_s \leq 100\,\mu m \\[3mm] \dfrac{10\upsilon}{d_s}\left[\left(1 + \dfrac{0.01(s-1)gd_s^3}{\upsilon^2}\right)^{0.5} - 1\right] & 100 < d_s < 1000\,\mu m \\[3mm] 1.1[(s-1)gd_s]^{0.5} & d_s \geq 1000\,\mu m \end{cases} \qquad (4\text{-}11)$$

in which d_s = sieve diameter.

The suspended sediment d_s generally is somewhat smaller than that of the bed depending on the composition of the bed and the strength of the flow dynamics. van Rijn (2007a)'s formula was employed to estimate the suspended sediment size,

$$d_s = \begin{cases} [1 + 0.006(d_{50}/d_{10} - 1)(\psi - 550)]d_{50} & \text{for} & \psi < 550 & \text{and} & d_{50} > d_{silt} \\ d_{50} & \text{for} & \psi \geq 550 & \text{and} & d_{50} > d_{silt} \\ d_{50} & \text{for} & d_{50} < d_{silt} \\ 0.5d_{silt} & \text{for} & d_{50} < 0.5d_{silt} \end{cases} \qquad (4\text{-}12)$$

where $\psi = u_{wc}^2 / [(s-1)gd_{50}]$ = mobility parameter. According to van Rijn's definition, $u_{wc}^2 = u_m^2 + u_c^2$, u_m = the peak orbital velocity, u_c = the depth-averaged current velocity, and $d_{silt} = 32$ µm. The lower limit is set to $d_{smin} = 0.5(d_{50} + d_{10})$.

Experiments have shown that the settling velocity is significantly reduced when the sediment concentration is high, which is the so-called hindered settling effect. For silt and sand, the hindered settling is slightly different, since the fluid movement around particles with $d < 100$ µm is laminar, and the fluid movement around settling particles with $d > 100$ µm is turbulent (Te Slaa et al., 2015). Silt and fine sand particles settle in the Stokes regime, and their geometry does not influence the hindered settling.

For sediment with grain size $d > 100 \mu m$, the settling velocity in a fluid-sediment suspension can be determined as (Richardson and Zaki, 1954; van Rijn, 1993):

$$w_s = w_{s,0}(1-c_v)^n \tag{4-13}$$

where w_s = the particle fall velocity, c_v = the volume sediment concentration of solids, and n is the exponent, varying from 4.6 to 2.3. The influence of the particle size on the hindered settling of sand is given by $n = 4.4(d_{50,ref} / d_{50})^{0.2}$, where $d_{50,ref}$ = 200 µm (Baldock et al., 2004).

A more generic hindered settling formula for silt and very fine sand was derived (Te Slaa et al., 2015):

$$w_s = w_{s,0} \frac{(1-c_v / \phi_{s,struct})^m (1-c_v)}{(1-c_v / \phi_{s,max})^{-2.5\phi_{s,max}}} \tag{4-14}$$

where $\phi_{s,struct}$ = 0.5 is the structural density, i.e., the solid content upon reaching the structural density of the bed, and $\phi_{s,max}$ = 0.65 is the maximum density, i.e., the solid content at the maximum packing of the particles. Upon reaching the structural density, a network of particles is formed and the settling velocities reduce to zero. m = 1-2 represents the effects of the return flow.

The formulae of hindered settling velocity come from settling column experiments. Under sheet flow conditions, near the bed level $z = 0$ in high-concentration area ($0.3 < c_v < 0.4$), Nielsen et al. (2002) found that the settling velocity is much lower than expected. We have to carefully choose the formulae of hindered settling velocity if study the sheet flow layer. This study does not penetrate into the sheet flow layer and focuses on the suspension layer above the reference height.

4.2.4.2. Sediment simulation related to bed forms

Ripples exhibiting the formation of fluid vortices (orbital excursion larger than ripple length) are called vortex ripples (Bagnold and Taylor, 1946). Hooshmand et al. (2015) suggested that for a silt-dominated sediment (d_{50} = 75 μm), Re_Δ = 450 (Stokes Reynolds number) is the critical condition between ripple dominated bed and non-rippled bed. Sheet flow occurs under stronger flow dynamics. Above plane beds, momentum transfer occurs primarily by turbulent diffusion; above rippled beds, momentum transfer and the associated sediment dynamics in the near-bed layer are dominated by coherent motions, especially the process of vortex formation above the ripple lee slopes and the shedding of these vortices at times of flow reversal (van der A, 2005).

1) Sheet flow conditions or 'flat bed'

The term 'flat bed' is used in this paper to refer to 'dynamically plane' rough beds, including sheet flow conditions and rippled beds of mild steepness (< 0.12) (Davies and Villaret, 2002), above which momentum transfer occurs via turbulent processes rather than vortices. Under these conditions, the normal k-ε turbulence model is solved, including the stratification effects.

The flow under sheet flow conditions is affected by the relatively thin sheet flow layer with high sediment concentration, such as the mobile bed effects (Nielsen, 1992). The enhanced roughness due to mobile bed effects has been studied by many scholars (e.g., Camenen et al., 2009; Dohmen-Janssen et al., 2001; Wilson, 1989). Camenen et al. (2009) proposed the Nikuradse's equivalent roughness by compiling many datasets,

$$\frac{k_s}{d_{50}} = 0.6 + 2.4\left(\frac{\theta}{\theta_{cr,ur}}\right)^{1.7} \qquad (4\text{-}15)$$

in which θ is the Shields parameter, and $\theta_{cr,ur} = 0.115F_{rw}^{1.2}/[W_{s*}^{0.4}(s-1)^{0.3}]$ is the critical Shields parameter for the inception of the upper regime. $F_{rw} = u_m/\sqrt{g\delta}$ is the wave Froude number, where u_m is the wave orbital velocity, $\delta = \sqrt{\upsilon T}$ is the thickness of the viscous (Stokes) layer, and T is the wave period. $W_{s*} = \left[(s-1)^2/(g\upsilon)\right]^{1/3} w_s$ is the dimensionless settling velocity. If $\theta < \theta_{cr,ur}$, then $k_s < 3d_{50}$, which corresponds approximately to the skin friction.

For finer sediments, the grain roughness becomes smaller and the enhanced roughness from mobile bed effects becomes more important.

2) Rippled bed

In a near-bed layer of approximately two ripple heights above the rippled bed, the flow dynamics are dominated by coherent periodic vortex structures, whereas above this layer the coherent motions break down and are replaced by random turbulence (Davies and Villaret, 1999). This leads to sediment in

73

suspension having considerably greater heights compared to flat beds. Ripple vortices are 3D or 2DV phenomena, and it is not physically justifiable to describe hydrodynamics and sediment dynamics over ripples with a 1DV approach. However, from a practical point of view, more sophisticated 2DV models are unduly complex and, therefore, 1DV models are preferred (van Der Werf, 2003). Recent research has proved the merits of the 1DV approach (Davies and Thorne, 2005; van der A, 2005; van der Werf et al., 2006). In the following sections, some key approaches are introduced regarding ripple prediction, roughness, vortex eddy viscosity and pick-up function.

a) Ripple prediction

There are many formulas for predicting ripple parameters (Khelifa and Ouellet, 2000; Mogridge et al., 1994; Nielsen, 1992). Khelifa and Ouellet (2000)'s method was employed in this study, which has been verified by many experimental and field data and can be used for both wave-only and combined wave-current conditions.

$$\frac{2\lambda}{d_{wc}} = 1.9 + 0.08 \ln^2(1 + \psi_{wck}) - 0.74 \ln(1 + \psi_{wck}) \qquad (4\text{-}16)$$

$$\frac{2\eta}{d_{wc}} = 0.32 + 0.017 \ln^2(1 + \psi_{wck}) - 0.142 \ln(1 + \psi_{wck}) \qquad (4\text{-}17)$$

where η is ripple height and λ is ripple length. $d_{wc} = T U_{wc}$, $\psi_{wck} = \dfrac{U_{wc}^{\,2}}{(s-1)gd}$,

$U_{wc} = \sqrt{(\dfrac{u_m}{\pi})^2 + \bar{u}_c^{\,2} + 2\dfrac{u_m \bar{u}_c}{\pi}|\cos\phi|}$, T = wave period, $\bar{u}_c$ = depth-averaged

current velocity, u_m = wave orbital velocity and ϕ = the angle between wave and current. Under wave and wave-current motions, the applicability of the formula has been tested for ψ_{wck} varying in the ranges of 0.3-20 and 0.7-145, respectively.

b) The bed form roughness is determined empirically from ripple parameters,

$$k_s = a_s \frac{\eta^2}{\lambda} \qquad (4\text{-}18)$$

where a_s is a constant. The factor a_s is still arguable, such as 8 (Nielsen, 1992), 25 (Davies and Thorne, 2005) and 27.7 (Grant and Madsen, 1982). Nielsen (1992) also suggested the roughness contribution from the moving sediment over ripples.

c) Eddy viscosity and diffusion coefficient

In accordance with the physical background, a two-layer model was adopted, i.e., the vortex-dominated layer at the bottom and the turbulence-dominated layer above, separated by twice the ripple height (Davies and Thorne, 2005; van der Werf et al., 2006).

In the vortex-dominated layer, the mean eddy viscosity adopts Nielsen (1992)'s height-invariant expression for very rough beds,

$$\overline{\upsilon_{tN}} = c_{vor} A \omega k_s \tag{4-19}$$

in which $c_{vor} = 0.004$-0.005.

The time varying eddy viscosity is assumed to be given by the real part of the following expression (Davies and Thorne, 2005)

$$\upsilon_{tN}(t) = \overline{\upsilon_{tN}} f(\omega t) \tag{4-20}$$

with $f(\omega t) = (1 + \varepsilon_0 + \varepsilon_1 e^{i\omega t} + \varepsilon_2 e^{2i\omega t})$. $f(\omega t)$ is briefly introduced in Appendix B. Please see Davies and Thorne (2005) for more details.

In the turbulence-dominated layer, the k-ε model was employed to provide the eddy viscosity. At the interface between the vortex-dominated layer and the turbulence-dominated layer, the values of k and ε were derived from the mixing length $l = 2\kappa\eta$ (Davies and Thorne, 2005) by $\varepsilon = c_\mu k^{3/2} / l$. Thus, at the edge of vortex layer,

$$k_{vortex}(t) = [\frac{\upsilon_{tN}(t)}{0.8\eta}]^2$$

$$\varepsilon_{vortex}(t) = c_\mu \frac{\upsilon_{tN}(t)^3}{(0.8\eta)^4} \tag{4-21}$$

Then above the vortex layer, the model reverts to the standard turbulence closure scheme.

The sediment diffusivity in the lower layer above rippled beds is significantly larger than the eddy viscosity (Nielsen, 1992; Thorne et al., 2002), with

$$\varepsilon_s(t) = \beta \frac{\upsilon_t(t)}{\sigma} \tag{4-22}$$

The value of the parameter has been assumed to revert smoothly from its value of $\beta = 4.0$ in the lower vortex layer towards unity in the upper layer, according to the power law rule:

$$\beta = 4.0 - 3.0(\frac{z - 2\eta}{h - 2\eta})^\gamma \tag{4-23}$$

where h is the water depth, z is the elevation from the bottom and the coefficient $\gamma = 0.4$-1.

d) Time-process of the reference concentration over rippled bed

The time-varying reference concentration is (Davies and Thorne, 2005)

$$c_a(t) = \overline{c_a} \frac{-0.5((1+\varepsilon_0)+\varepsilon_1 e^{i\omega t}+\varepsilon_2 e^{2i\omega t})((1+a_c e^{2i\omega t})+c.c)}{((1+\varepsilon_0)+0.25 A_c |\varepsilon_0|(e^{i(2\phi_1-2\phi_c)}+c.c))} \qquad (4\text{-}24)$$

in which $\overline{c_a}$ is the mean reference sediment concentration. Eq. (4-24) is briefly introduced in Appendix B. For more details, please see Davies and Thorne (2005).

4.2.4.3. Stratification effects

When sediment is suspended, the vertical gradient of sediment concentration causes the vertical gradient of density to increase. Studies reported in literatures show an appreciable effect of suspended sediment on the turbulence properties (Traykovski et al., 2007; Winterwerp, 1999). If the sediment concentration gradient is high, sediment-induced turbulence damping can largely affect the velocity profile and the transport rate, especially for fine sediment (Conley et al., 2008; Hassan and Ribberink, 2010; Kranenburg et al., 2013; Winterwerp, 2001). Thus, the sediment-induced turbulence damping is an important term for high concentration layer modelling. Some models consider sediment-flow interaction processes in different ways (Hassan and Ribberink, 2010; Kranenburg et al., 2013). Generally, the buoyancy flux B_k accounts for the conversion of turbulent kinetic energy to mean potential energy with the mixing of sediment, which is considered equivalent to buoyancy flux in a salt-stratified or thermally stratified flow. The following expressions are used to describe the buoyancy flux B_k and the Brunt-Vaisala frequency N,

$$B_k = \frac{\upsilon_t}{\sigma_v} N^2; \qquad N^2 = -\frac{g}{\rho_m}\frac{\partial \rho_m}{\partial z} \qquad (4\text{-}25)$$

The coefficient in Eq. (4-5) $c_{3\varepsilon}=0 \quad N^2>0$ and $c_{3\varepsilon}=1 \quad N^2<0$. ρ_m is the density of the local water-sediment mixture,

$$\rho_m = \rho+(\rho_s-\rho)c_v \qquad (4\text{-}26)$$

in which ρ is the density of clear water, ρ_s is the sediment particle density, and c_v is the volumetric sediment concentration.

To evaluate the damping of vortex viscosity, van Rijn (2007a)'s formula for damping effect was employed by a damping coefficient ϕ_d,

$$\upsilon_{tN}(t) = \phi_d \upsilon_{tN}(t) \qquad (4\text{-}27)$$

in which $\phi_d = \phi_{fs}[1+(c/c_{gel,s})^{0.8}-2(c/c_{gel,s})^{0.4}]$, $\phi_{fs}=d_{50}/(1.5 d_{sand})$, and $\phi_{fs}=1$ for $d_{50} \geq 1.5 d_{sand}$. $c_{gel,s}=0.65$ = maximum bed concentration in volume.

4.2.4.4. Reference concentration

(1) Approaches for reference concentration

The reference concentration considering silt was employed (Yao et al., 2015).

$$c_a = \beta_y (1 - p_{clay}) f_{silt} \frac{d_{50}}{z_a} \frac{T_*^{1.5}}{D_*^{0.3}} \tag{4-28}$$

in which $\beta_y = 0.015$ is an original empirical coefficient for sand, and Yao et al. (2015) extended it to silt by using $\beta_y = 0.118 D_*^{-0.7}$, with a maximum value of 0.118 and minimum value of 0.015. $f_{silt} = d_{sand}/d_{50}$ is the silt factor ($f_{silt} = 1$ for $d_{50} > d_{sand}$), and d_{sand} = 62 μm. p_{clay} is the percentage of clay material in the bed. $D_* = d_{50}[(s-1)g/\upsilon^2]^{1/3}$ is the dimensionless particle size. $T_* = (\tau'-\tau_c)/\tau_c$, in which τ' is originally the time-averaged effective bed-shear stress under currents and waves. τ_{cr} is the critical bed shear stress. The reference height z_a is defined as the maximum value of half the wave-related and half the current-related bed roughness values, with a minimum value of 0.01 m.

For sand simulation in sheet flow condition and rippled beds, Zyserman and Fredsøe (1994)'s formula and Nielsen (1992)'s formula were recommended respectively.

Zyserman and Fredsøe (1994)'s formula:

$$c_a = \frac{0.331(\theta-\theta_c)^{1.75}}{1+0.720(\theta-\theta_c)^{1.75}}, \text{ at } z_a = 2d_{50} \tag{4-29}$$

in which θ is the instantaneous Shields number and θ_c is the critical Shields number.

Nielsen (1992)'s formula:

$$\overline{c_a} = 0.0022\theta_r^3, \text{ at } z_a = 2d_{50} \tag{4-30}$$

in which θ_r is the ripple-adjusted value of Shields number.

When applying the formulas of the reference concentration in oscillatory flows, a zero value for the bed concentration is unrealistically obtained during the stage of a wave cycle, when the shear stress is lower than the critical value (Fredsøe and Deigaard, 1992). To overcome this shortcoming, the deposited sediment from the last time step was considered here. At the reference level, from the governing equation, the diffusion is ignored:

$$\frac{\partial c}{\partial t} - w_s \frac{\partial c}{\partial z} = 0.$$

$$c_a^n = c_a^{n-1} + w_s \frac{\Delta t}{\Delta z}(c_{za+1}^{n-1} - c_a^{n-1}) \tag{4-31}$$

Then, $c_a(t) = \max(c_a^n, c_a)$, in which, c_a^n is the reference sediment concentration at the present time step, c_{za+1}^{n-1} is the c at the grid above the reference height z_a at the last time step, and c_a^{n-1} is the reference sediment concentration at the last time step. Figure 4-1 shows the comparison of the non-adjusted and adjusted reference sediment concentration.

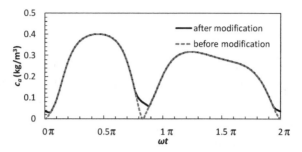

Figure 4-1. *Comparison of reference concentration at the reference height and the revised version (in case of the experiment conditions of O'Donoghue and Wright (2004) (the black line represents the revised reference concentration and the red dash line represents the original reference concentration)*

(2) Critical shear stress of sediment incipience motion

In the above formulas, the critical shear stress needs to be determined and generally the Shields curve can be employed. However, the Shields curve which is normally used for non-cohesive sediments cannot be used for silt. The expression of silt-sand incipience motion proposed in chapter 3 was employed here, which considered the cohesive force and additional static water pressure for fine sediment (Zuo et al., 2017).

$$\theta_{zc} = \begin{cases} 0.025\,\mathrm{Re}\,d_* ^{-0.07} & \mathrm{Re}\,d_* < 1 \\ 0.00543\ln(\mathrm{Re}\,d_*) + 0.025 & 1 \le \mathrm{Re}\,d_* \le 100 \\ 0.05 & \mathrm{Re}\,d_* > 100 \end{cases} \qquad (4\text{-}32)$$

where $\theta_{zc} = \dfrac{\tau_c}{\rho(s-1)gd + a\beta_c\rho\dfrac{\varepsilon_k + gh\delta_s\sqrt{\delta_s/d}}{d}}$; $\mathrm{Re}\,d_* = \dfrac{d}{4\upsilon}\sqrt{(s-1)gd}$ is the

non-dimensional sand Reynolds number; τ_c is the critical shear stress; $\varepsilon_k = 1.75\times10^{-6}$ m³/s² is the cohesive force coefficient; $\delta_s = 2.31\times10^{-7}$ m is the bound water thickness; $a = 0.19$ is a coefficient, and β_c is the compaction coefficient, normally $\beta_c = 1$ for well-compacted sediments.

4.2.5. Boundary conditions and initial conditions

At the bottom: $u(z_0,t) = 0$, $w(z_0,t) = 0$ $\qquad z_0 = k_s/30$

$$k(z_0,t) = v_t \left| \frac{\partial u}{\partial z} \right| / \sqrt{c_1} \ , \ \varepsilon(z_0,t) = (c_1)^{3/4} \frac{k^{3/2}}{\kappa z_0}$$

$$\varepsilon_s \frac{\partial c}{\partial z} = -\omega_s c_a \qquad\qquad \text{at } z = z_a$$

At the upper boundary: For waves alone, the upper edge of the flow domain, $z = z_{max}$ is chosen where the boundary layer effects have disappeared. The condition of no shear is applied at the edge of the bottom boundary layer at $z = z_{max}$. The Neumann condition is applied on the velocity,

$$\frac{\partial}{\partial z} u(z_{max}, t) = 0.$$

Zero flux conditions are imposed for the turbulent quantities at the edge of the flow domain,

$$\frac{\partial k}{\partial z} = \frac{\partial \varepsilon}{\partial z} = 0.$$

The velocity at z_{max} (upper boundary) is given by the Dirichlet condition
$$u(z_{max}, t) = \bar{u} + u_\infty(t),$$

in which $\bar{u}$ is the mean flow velocity, and $u_\infty(t)$ is the wave orbital free stream velocity.

A zero flux condition is imposed on the sediment concentration at the upper boundary:

$$(\omega_s - w)c + \varepsilon_s \frac{\partial c}{\partial z} = 0 \qquad \text{at } z = z_{max}$$

4.2.6. Numerical discretization

Geometric stretching of the mesh was applied to obtain a fine resolution close to the bed (Zhang et al., 2011), and a stretching factor of 1.05 was applied. Our experience shows that the grid structure with 50-100 vertical grid cells is sufficient for resolving the boundary layer in the following cases.

The FVM (finite volume method) method was employed to discretize the governing equations. Time discretization is based on the θ_f method. A coefficient θ_f was employed. When θ_f = 0, the discretization scheme is explicit; when θ_f = 1, it is an implicit scheme; and when θ_f = 0.5, it is a Crank-Nicholson semi-implicit scheme. In this study, the implicit scheme was employed. The convection term was discretized by a first-order upwind scheme. After discretization, tridiagonal matrices were obtained, and the TDMA method (tridiagonal matrix algorithm) was employed to solve the matrices. The convergence condition was settled as 10^{-5} for the u, v_t, k, ε and c at the same phase between the two periods.

4.3. Model verification

4.3.1. Flow simulation in wave-current BBL

Experimental data of Jensen et al. (1989), Klopman (1994), and Umeyama (2005) were used to verify the model. Figure 4-2 to Figure 4-5 show the verification of Jensen et al. (1989)'s experiment data (Test 13). The experiments were carried out in a U-shaped oscillatory-flow water tunnel. The velocity amplitude was 2 m/s and the wave period was 9.72 s. A sheet of sandpaper was glued at the bottom with a sand-roughness value of k_s = 0.84 mm. The oscillatory flow in the tunnel was driven by an electronically controlled pneumatic system, and the velocity was sinusoidal. For more details, please refer to Jensen et al. (1989). According to Tanaka and Thu (1994), the laminar wave boundary was defined as Reynolds number $Re_{wave} = Au_m / \upsilon$ less than 1.8×10^5. The wave Reynolds number in this case was 61.9×10^5, which indicates that it was a turbulent wave boundary layer and the standard $k\text{-}\varepsilon$ model was adopted here. The simulated velocity distribution in different phases matched the measured data well. The simulated velocity defect, kinetic energy, shear stress and shear velocity matched the measured data.

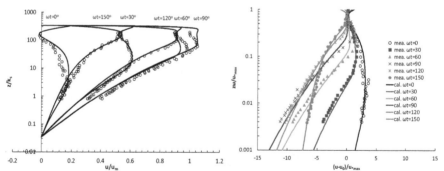

Figure 4-2. *Comparison of calculated and measured wave velocity profiles (Jensen et al. (1989)'s experiment, dots were measured, lines were calculated and ω is the angular frequency)*

Figure 4-3. *Verification of velocity defect of test 13 of Jensen et al. (1989)'s experiment*

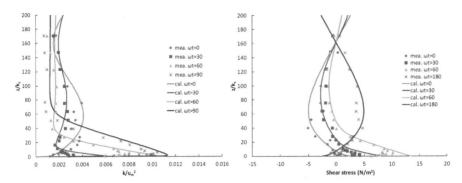

Figure 4-4. *Verification of kinetic energy distribution of test 13 of Jensen et al. (1989)'s experiment*

Figure 4-5. *Verification of shear stress distribution of test 13 of Jensen et al. (1989)'s experiment*

Figure 4-6 and Figure 4-7 show the velocity amplitude, the mean velocity distribution under the wave-only condition and the combined wave-current condition in Klopman (1994)'s experiment. The wave height was 0.12 m, wave period was 1.4 s and k_s was 0.84 mm. The mean current velocity was 0.16 m/s and water depth was 0.5 m. It can be seen that the model simulated the velocity distribution well. In this case, the wave Reynolds number is 5.9×10^3, which means that it is a laminar wave boundary layer. The NTM turbulence model could get a better result, while the standard k-ε turbulence model overestimated the BBL (Figure 4-6). In the wave bottom boundary layer, beyond the log-distribution layer, there was an over-shoot at the edge of the BBL. The wave-induced current was positive near the bottom and negative above the boundary layer. For the wave following current, the velocity near the bottom mainly increases compared with the current-only case, while for the wave opposing current, it decreases. At the upper part, the change of velocity show a different tendency related to Stokes drift effects, which is not shown in this paper.

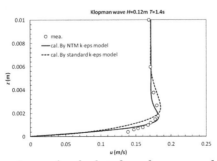

Figure 4-6. *Comparison of calculated and measured velocity amplitude*

profile of wave-only case (Klopman (1994)'s experiment)

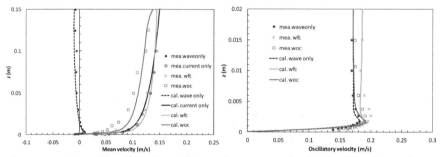

Figure 4-7. *Verification of profiles of the mean velocity (left) and oscillatory velocity (right) in Klopman (1994)'s experiment (wfc: wave following current; woc: wave opposing current)*

Umeyama (2005) carried out experiments to study the changes in the mean velocity profile owing to the interaction between waves and currents in a recirculating wave tank. The wave tank is 25 m long, 0.7 m wide and 1.0 m deep. Figure 4-8 shows the comparison of the measured and calculated phase-average velocity distribution for the wave-only case (W2, W4), wave following case (WCF2, WCF4) and wave opposing case (WCA2, WCA4). The results showed that the calculated mean velocity for different cases matched the measured data well.

Table 4-2. *Experiment conditions of Umeyama (2005)'s experiment*

Cases	h (m)	H (m)	T (s)	u_c (m/s)	$\bar{u}$ at z (m/s)	z (m)	note
W2	0.2	0.0251	1.0	0	0	0.2	Wave only
W4	0.2	0.028	1.4	0	0	0.2	Wave only
WCF2	0.2	0.0231	1.0	0.12	0.111	0.05	Wave following current
WCF4	0.2	0.0250	1.4	0.12	0.117	0.05	Wave following current
WCA2	0.2	0.0215	1.0	-0.12	0.112	0.07	Wave opposing current
WCA4	0.2	0.0270	1.4	-0.12	0.118	0.07	Wave opposing current

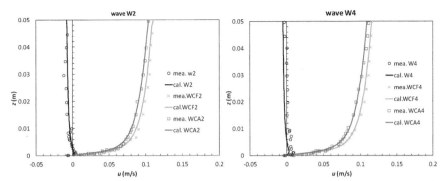

Figure 4-8. *Verification of mean velocity distribution of Umeyama (2005)'s experiment in case of wave-only (W2, W4), wave following current (WCF2, WCF4) and wave against current (WCA2, WCA4)*

Based on the validation, it can be concluded that the model is able to simulate the flow dynamics near the bed bottom in wave-current conditions. It is able to simulate the mean velocity distribution, velocity amplitude, Reynolds stress, eddy viscosity, turbulence and phase defect within the wave period. This model has been implemented and validated not only for the full turbulence case, but also for low-Reynolds number cases. It can simulate the wave-induced net current and the combined wave-current interaction. All these processes are important for sediment transport in the BBL, which means that this model can provide suitable flow dynamics for sediment simulation. However, this model is still not able to simulate the velocity distribution near the surface as the Stokes drift is not taken into account. This is acceptable for the purpose of studying the high sediment concentration near the bottom.

4.3.2. Verification of sediment movement in the wave-current BBL

Experimental data of Horikawa et al. (1982), Ribberink and Al-Salem (1995), O'Donoghue and Wright (2004), Katopodi et al. (1994), Dohmen-Janssen et al. (2001), Williams et al. (1998), Li (2014), Zhou and Ju (2007) and Yao et al. (2015) were used to verify the model, as listed in Table 4-3. The bed forms included sheet flow and rippled bed; the flow dynamics conditions included wave only cases and combined wave-current cases; and the sediment materials included silt and sand. There were still few experiments on silty sediments under sheet flow conditions, thus we collected some experimental datasets on fine sand to verify the model. Besides, in Yao et al. (2015)'s experiment on silt, the case of s1-f3212 and s1-o3812, ripples disappeared when currents were imposed, which could present an evidence of 'flat bed'.

4.3.2.1. Sheet flow cases

(a) Case of Ribberink and Al-Salem (1995)

Ribberink and Al-Salem (1995) carried out experiments under sheet flow conditions, using an oscillating water tunnel. The sand bed in the test section consisted of quartz sand with d_{50} of 0.21 mm and d_{10} of 0.15 mm. Figure 4-9 shows the mean sediment concentration profile resulting from a sinusoidal wave with a velocity amplitude of 1.7 m/s and a wave period of 7.2 s (case C3). The predicted mean sediment concentration profile was in good agreement with the measurements. Figure 4-9 also shows the variation of the sediment concentration through the wave cycle at 0.5 cm, 1.1 cm and 2.1 cm above the bed. It can be seen that, although there were some deviations in the magnitude during the intra-wave process, the variation of sediment concentration through the wave period matched the measured data. The calculated sediment concentration near bottom was in accordance with the imposed flow dynamics, while the measured data showed a phase lag. This is mainly because the instantaneous reference concentration was calculated based on shear stress; however, a sheet flow layer exists below the reference height, and the sheet flow layer would affect the above sediment concentration. According to Nielsen (2002), the concentration at the undisturbed bed level varies very little with time but the sediment flux varies strongly. To describe this process, a two-phase flow model would be needed, which could penetrate into the sheet flow layer.

Table 4-3. *Verification cases for sediment movenment*

Cases	Flow dynamics	Wave type	d_{50} (mm)	Bed forms
Horikawa et al. (1982)	Wave only	Oscillatory tunnel	0.20	Sheet flow
Ribberink and Al-Salem (1995)	Wave only	Oscillatory tunnel	0.21	Sheet flow
O'Donoghue and Wright (2004)	Wave only	Oscillatory tunnel	0.15-0.51	Sheet flow
Katopodi et al. (1994)	Wave+current	Oscillatory tunnel	0.21	Sheet flow
Dohmen-Janssen et al. (2001)	Wave+current	Wave flume	0.13-0.32	Sheet flow
Williams et al. (1998)	Wave only	Wave flume	0.329	Rippled bed
Li (2014)	Wave only	Wave flume	0.045-0.11	Rippled bed
Zhou and Ju (2007)	Wave+current	Wave flume	0.062-0.11	Rippled bed
Yao et al. (2015)	Wave and wave+current	Wave flume	0.046-0.0 88	Rippled bed & flat bed

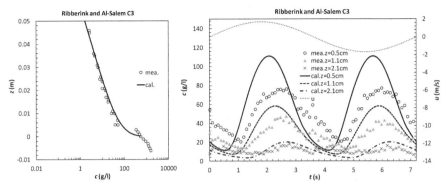

Figure 4-9. *Calculated and measured mean sediment concentration beneath sinusoidal waves (left: mean sediment concentration profile; right: sediment concentrations through the wave cycle at different elevations above the bed)*

(b) Case of O'Donoghue and Wright (2004)

O'Donoghue and Wright (2004) carried out experiments in an oscillatory tunnel under asymmetry oscillatory motion, which was described as $u_\infty(t) = U_1 \sin\omega t - U_2 \cos 2\omega t$, with U_1 = 1.2 m/s and U_2 = 0.3 m/s, T = 5 s, see Figure 4-10. The experimental sediments comprised three well-sorted sands, fine, medium and coarse sand with d_{50} = 0.15, 0.28 and 0.51 mm, d_{10} = 0.10, 0.17 and 0.36 mm, respectively.

Figure 4-11 shows the comparison of the calculated and measured values of the mean sediment concentration profiles. It can be concluded that, the model is able to simulate the sediment concentration profile changes as results of the sediment grain size changes. Figure 4-12 and Figure 4-13 show the sediment flux profiles in different phases during the intra-wave process. The calculated flux profiles in most phases were in agreement with the measured ones. As expected, the flux magnitudes were much larger in the case of the fine sand compared with the medium sand. This was due to the higher suspended concentrations in the case of fine sand.

Figure 4-14 shows the comparison of the calculated and measured time-averaged flux profiles. Although deviations exist in the values, the model is able to simulate the changes of the flux direction for different grain particles, i.e., the flux of 'fine' sand (FA5010) shows offshore direction, while the fluxes of 'medium' and 'coarse' sands (MA5010 and CA5010) show onshore direction. However, in the case of fine sand, the calculated fluxes in the minus flow phase (t/T = 0.42-0.71 in case FA5010) were not as well simulated. Although the maximum offshore velocity was only approximately 60% of the maximum onshore velocity, the flux value in the phase of maximum offshore velocity (t/T

= 0.71) was similar to that at the maximum onshore velocity (t/T = 0.21). This might have been caused by the unsteady effects that occurred in the case of fine sand (Dohmen-Janssen et al., 2002). As investigated during the experiment O'Donoghue and Wright (2004), at t/T = 0.21, the fine sand was carried high into the flow as a result of the high flow velocities, contributing to the high onshore flux at this phase. Because of its low settling velocity, the fine sand slowly settled, however, a significant proportion did not settle back to the bed as the flow velocity decreased. The high offshore flux during the offshore flow was therefore caused by the presence of the high concentrations resulting from the slow settling of sediment entrained by the previous high onshore velocities.

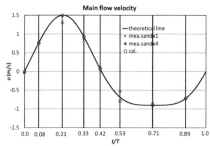 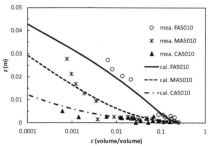

Figure 4-10. *Time series of the main flow velocity in O'Donoghue and Wright (2004)'s experiment*

Figure 4-11. *Verification of mean sediment concentration profiles for different sand sediments of O'Donoghue and Wright (2004)'s experiment*

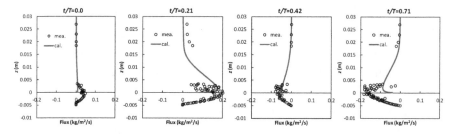

Figure 4-12. *Verification of flux profiles at selected phases for experiment FA5010 (values below 0 are in sheet flow layer which are not included in the model)*

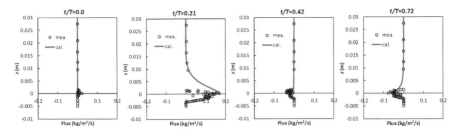

Figure 4-13. *Verification of flux profiles at selected phases for experiment MA5010 (values below 0 are in sheet flow layer which are not included in the model)*

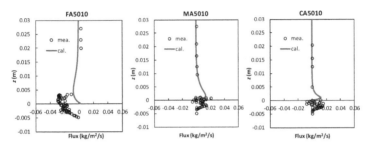

Figure 4-14. *Comparison of calculated and measured time-averaged flux profiles*

(c) Dohmen-Janssen et al. (2001)'s case

Dohmen-Janssen et al. (2001)'s experiments were carried out in the Large Oscillating Water Tunnel (LOWT) of Delft Hydraulics (now Deltares), in which near-bed orbital velocities in combination with a net current could be simulated at full scale. Three uniform sands with different mean grain sizes were used. The experimental conditions were listed in Table 4-4. Figure 4-15 shows the comparison of the measured and calculated sediment concentration profiles. It can be concluded that, the model is able to simulate the mean concentration profiles for different grain sizes.

Table 4-4. *Experimental conditions of Dohmen-Janssen et al. (2001)*

Case	h (m)	u_m (m)	T (s)	u_c (m/s)	d_{50} (mm)	d_{10} (mm)
D1	0.8	1.47	7.2	0.24	0.13	0.10
D2	0.8	1.47	7.2	0.23	0.21	0.15
D3	0.8	1.47	7.2	0.26	0.32	0.22

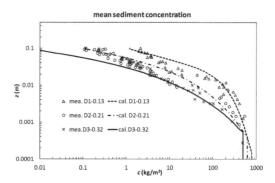

Figure 4-15. *Comparison of measured and calculated sediment concentration profiles of Dohmen-Janssen et al. (2001)*

(d) Horikawa et al. (1982)'s case

Horikawa et al. (1982) measured instantaneous sediment concentration and velocity under sinusoidal waves in an oscillating water tunnel. The median grain diameter was 0.2 mm. The wave excursion amplitude was held constant at 0.72 m, while the wave period varied from 3.6 to 6.0 s. Figure 4-16 shows the mean vertical profiles of suspended sediment concentration for a range of wave parameters. Figure 4-17 shows the verification of instantaneous sediment concentration, velocity and sediment flux at different phases. The calculated value is in fair agreement with the measured results.

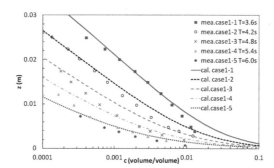

Figure 4-16. *Verification of mean sediment concentration profiles of Horikawa et al. (1982)'s experiment*

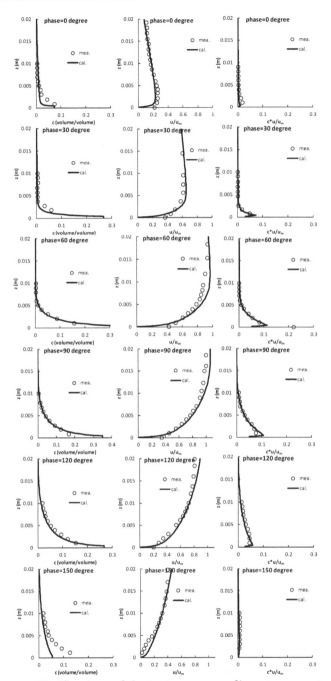

Figure 4-17. *Verification of instantaneous sediment concentration (left column), velocity (middle column) and sediment flux (right column) at different phases (Horikawa et al. (1982)'s experiment, case1-1, A = 0.72 m, T = 3.6s, u_m = 1.27m/s)*

(e) Verification of Katopodi et al. (1994)'s experiment, and the measured data was reproduced from Li and Davies (1996)

The experiments were carried out in a large oscillatory water tunnel (14 m in length, 1.1 m in height and 0.3 m in width) at Delft Hydraulics. In the wave-current (series E) experiments, measurements of the time-dependent velocity and sediment concentration were made at various heights above the bed for four sinusoidal wave and current combinations (coded E1, E2, E3, E4) under sheet flow conditions. It can be seen that the verifications of the mean velocity, mean sediment concentration and sediment mass flux were fairly well, as well as the time-averaged sediment mass flux (Figure 4-18 and Figure 4-19). Figure 4-20 shows the comparison of the measured and calculated time-dependent sediment mass concentration during one wave period at levels $z = 1.45, 2.35, 3.65, 5.35$ cm for experiment E1.

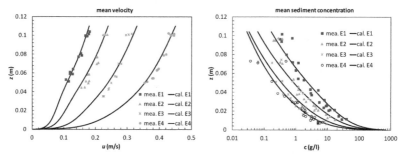

Figure 4-18. Comparison of the calculated and the measured mean velocity profiles (left) and mean suspended sediment concentration profiles (right)

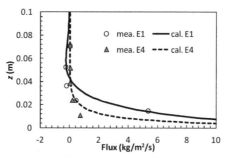

Figure 4-19. Comparison of the measured (symbols) and the computed (lines) cycle-averaged sediment mass flux for experiments E1 and E4

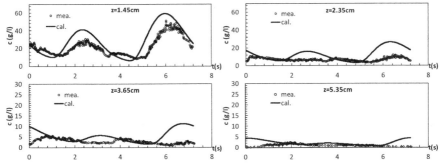

Figure 4-20. *Comparison of the measured and the calculated time-dependent sediment mass concentration during one wave period at levels z = 1.45, 2.35, 3.65, 5.35 cm for experiment E1, corresponding with time-dependent horizontal velocity at level z = 7.75 cm*

4.3.2.2. Rippled bed case

Ripples can form in sand bed (Williams et al., 1998) and silt bed (Zhou and Ju, 2007; Yao et al., 2015). The experimental datasets were used to verify the model.

(a) Williams et al. (1998)'s case

Detailed measurements of sediment in suspension above rippled bed have been made in the Deltaflume of Delft Hydraulics (now Deltares). Table 4-5 shows the experiment conditions. A sediment bed, 0.5 m thick and 30 m long, was placed approximately halfway along the flume, above which the water depth was 4.5 m in each test. The bed sediment comprised sand with median diameter d_{50} = 0.329 mm and d_{10} = 0.175 mm. Figure 4-21 shows the verification of mean sediment concentration profile. It can be seen that the model is able to simulate the mean concentration profile reasonably. Figure 4-22 shows the time-dependent profiles of the turbulent kinetic energy and eddy viscosity in an intra-wave process. The vortex viscosity dominated the near bottom layer and the turbulence developed in the upper layer.

Table 4-5. *Experimental conditions of Williams et al. (1998)*

Case	U_1 (m/s)	U_2 (m/s)	T(s)	Ripple height (m)	Ripple length (m)	d_{50} (mm)	d_{10} (mm)	w_s (mm/s)
Test 4	0.652	0.043	5	0.065	0.51	0.329	0.175	29.4
Test 6	0.534	0.0288	5	0.059	0.42	0.329	0.175	24.7

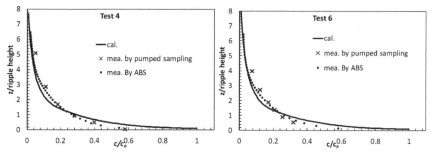

Figure 4-21. *Comparison of measured and calculated sediment concentration profile*

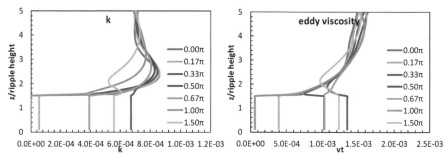

Figure 4-22. *Time-dependent profiles of turbulent kinetic energy (left) and eddy viscosity (right) (Note: As the measured data were collected above the ripple crest, the initial bed level was set at the ripple crest, and thus the vortex layer height was 1.5 times the ripple height showing in the figures)*

(b) Zhou and Ju (2007)'s experiment

Zhou and Ju (2007)'s experiment was carried out in a wave flume in Nanjing Hydraulic Research Institute. The flume is 175 m long, 1.2 m wide and 1.6 m deep. The experimental sediments are fine sediments with d_{50} = 0.062 mm and 0.11 mm, water depth h = 50 cm, wave period T = 2 s, wave height H = 0.1-0.2 m, and mean current velocity u_c = 0.0, 0.123, 0.188, 0.253, 0.319 m/s.

Figure 4-23 and Figure 4-24 show the measured and calculated sediment concentration profiles under wave-only and wave-current conditions. According to Khelifa and Ouellet (2000)'s formula, for d_{50} = 0.062 mm and d_{50} = 0.11 m, the calculated ripple height is 0.5-0.7 cm and 1.0-1.2 cm, respectively, and the ripple length is 4.7-5.1 cm and 7.8-8.5 cm, respectively. Under the wave-only conditions, the measured sediment concentration profile can be considered as a fully developed equilibrium profile because of the relatively small net current. However, when the current is added, sufficient sediment source and a certain distance are needed to establish the equilibrium concentration. The length of sediment section in this experiment was only 10 m, and it was too short to develop the equilibrium concentration for combined wave-current conditions.

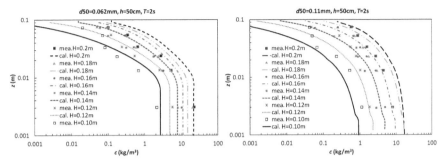

Figure 4-23. *Comparison of measured and calculated sediment concentration profiles under wave-only conditions*

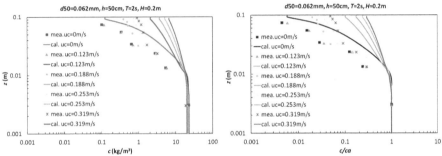

Figure 4-24. *Comparison of measured and calculated sediment concentration profiles under combined wave-current conditions (left: sediment mass; right: ratio of sediment concentration to reference concentration)*

Under the wave-only conditions, the calculated sediment concentration agreed well with the measured data; while under the combined wave-current conditions, the calculated sediment concentration (equilibrium) was larger than the measured value (non-equilibrium), which could be explained by the un-fully developed sediment concentration during the experiments. The non-equilibrium concentration can be further simulated by a 2DV model considering longitudinal diffusive transport.

However, despite the discrepancies between the computed and measured data, the model was able to simulate that the sediment profile became straighter as the current velocity increased. In the lower part, below about two times of the ripple height, the profiles changed little which indicated that waves dominated the sediment suspension near the bottom; while in the upper part, the concentration increased with the increase of the current velocity, which indicated that the currents dominated the sediment suspension in the upper part.

(c) Yao et al. (2015)'s experiment

Yao et al. (2015) conducted a series of flume experiments to investigate sediment transport of sand-silt mixtures in both wave-only and wave-current conditions. Two types of sediments were used: a silt-sized mixture with a median grain size of 46 μm, and a very fine sand-sized mixture with a median grain size of 88 μm. The experiment conditions are listed in Table 4-6. Figure 4-25 and Figure 4-26 show the measured and calculated sediment concentration profiles under wave-only and wave-current conditions. It can be seen that the calculated sediment concentration fits well with the measured data in many cases. There is larger discrepancy for s1 sediment, likely because larger mixture of sediments and two-layer bed morphology were detected during the experiments, i.e., silt-dominant and sand-dominant layers.

Table 4-6. *Experimental conditions of Yao et al. (2015)*

Case	h (m)	H (m)	T (s)	u_c (m/s)	Ripple height (cm)	Ripple length (cm)	Bed forms
S1-09	0.30	0.088	1.5	0	0.88	5.09	Rippled bed
S1-11	0.30	0.106	1.5	0	0.77	4.95	Rippled bed
S1-13	0.30	0.133	1.5	0	0.66	4.82	Rippled bed
S1-f3212	0.30	0.115	1.5	0.32	-	-	Flat bed
S1-03812	0.30	0.12	1.5	-0.38	-	-	Flat bed
S2-09	0.30	0.091	1.5	0	0.83	5.48	Rippled bed
S2-10	0.30	0.10	1.5	0	0.75	4.73	Rippled bed
S2-12	0.30	0.12	1.5	0	0.8	6.13	Rippled bed
S2-f3311	0.30	0.106	1.5	0.33	1.58	8.45	Rippled bed
S2-03911	0.30	0.11	1.5	-0.39	1.34	8.66	Rippled bed

Under combined wave-current conditions, the sediment diffused to the upper part and the sediment concentration profile was much straighter, which was similar to Zhou and Ju (2007)'s experiment. The length of the sediment section in Yao et al. (2015)'s experiment was 15 m, which was longer than that in Zhou and Ju (2007)'s experiment. However, this length is still not long enough to avoid the occurrence of non-equilibrium sediment concentration under the wave-current condition (Yao et al., 2015). Furthermore, for s1-f3212 and s1-03812 case, when currents were imposed, ripples disappeared as presented by the author, and the sediment concentration decreased without the vortexes' effects compared with that of s1-09~s1-13. It indicates that the model is able to simulate sediment concentration distribution relating to bed forms.

During simulation, the revised van Rijn (2007b)'s formula by Yao et al. (2015) (Eq. (4-28)) was employed for reference concentration. This formula is mainly for rippled bed. In this paper, however, the time-variant bed-shear stress with stratification effect was also used for silt simulation over 'flat bed' and showed reasonable results.

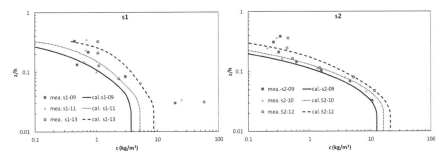

Figure 4-25. *Comparison of calculated and measured sediment concentration profile under wave-only cases of Yao et al. (2015)'s experiment*

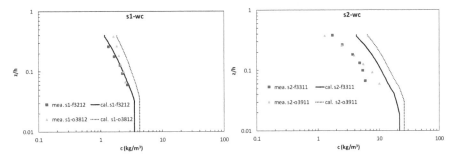

Figure 4-26. *Comparison of calculated and measured sediment concentration profiles under wave-current cases of Yao et al. (2015)'s experiment (left: flat bed; right: rippled bed)*

(d) Verification of Li (2014)'s flume wave case

Li (2014)'s experimental data was employed to verify the model over rippled bed. The sediment was silt and fine sand with d_{50} of 0.045 mm and 0.11 mm. The experimental wave heights were 0.12 m, 0.15 m, 0.18 m, 0.21 m, and the water depth was 0.5 m. Figure 4-27 shows the comparison of the measured and the calculated mean sediment profiles under different wave conditions as well as different sediment grain sizes. It can be seen that the sediment concentration near the bottom increased when the wave height increased. The sediment concentration profiles under waves were reasonably simulated by the model.

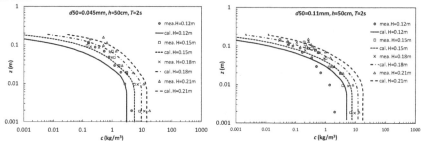

Figure 4-27. *Comparison of the measured and the calculated sediment concentration profiles (Li (2014)'s experiment, left: d_{50} = 0.045 mm; right: d_{50} = 0.11 mm)*

4.4. Sensitivity analysis and discussion: Factors that impact the sediment concentration profile of the HCL

Based on the validated experimental data, it is clear that a high concentration layer usually develops near the bottom under wave-dominant conditions. As sediment suspension occurs due to the turbulence diffusivity, the HCL is affected by the wave BBL. Over different bed forms, the eddy viscosity distribution as well as the sediment diffusion is different. Stratification effects may be an important impact factor on the HCL and the question is whether the stratification brings the collapse of turbulence. Hindered settling of silt has been studied extensively by Te Slaa et al. (2015) and the results clearly showed a large influence on the sediment concentration profile, thus the hindered settling is not discussed in this paper. Sensitivity analysis was carried out using the 1DV model (the calculation conditions were listed in Table 4-7), focused on the factors that might influence the HCL:

- The relation of the HCL and the wave BBL;
- Effects of bed forms: vortex diffusion induced by ripples and eddy diffusion of flat bed (sheet flow); and
- Effects of stratification and mobile bed roughness.

4.4.1. Relation of the HCL and the BBL

(1) Thickness of the HCL and the wave BBL

The formation of the HCL is strongly related to the turbulence production inside the wave boundary layer. For silt with a diameter of 62 μm, calculations were carried out with increasing orbital velocity and mobility number under oscillatory wave motions. Figure 4-28 shows the mean sediment concentration profiles and maximum orbital velocity profiles near the bed. With the increasing of the mobility number, the bed forms of the study cases change from rippled bed to flat-bed (sheet flow). Figure 4-29 shows the eddy viscosity distribution under different wave conditions, and Figure 4-30 shows the ripple parameters. In this study, the height of the HCL is defined where the gradient of the sediment concentration changes abruptly. For ripple cases d1-1 to d1-6, the HCL height varies at about 1.3-2.2 cm, and the corresponding wave boundary layer thickness is about 0.6-1.1 cm. For sheet flow cases d1-7 to d1-10, the HCL height varies at about 1.2-1.5 cm, and the corresponding wave boundary layer thickness is about 0.55-0.70 cm. The thickness of the HCL is about twice the wave boundary layer, as referred in some literature (Yao et al., 2015). It means that, although the velocity is restricted in the BBL, the eddy viscosity as well as the diffusion viscosity can still reach higher levels.

Table 4-7. *Calculation conditions of sensitivity study*

Case	h (m)	u_m (m/s)	T (s)	u_c (m/s)	Mobility number	d_{50} (mm)	Bed types
d1-1	0.3	0.12	3	0	14.3	0.062	ripple
d1-2	0.3	0.20	3	0	39.8	0.062	ripple
d1-3	0.3	0.25	3	0	62.2	0.062	ripple
d1-4	0.3	0.30	3	0	89.7	0.062	ripple
d1-5	0.3	0.35	3	0	122.1	0.062	ripple
d1-6	0.3	0.38	3	0	143.9	0.062	ripple
d1-7	0.3	0.50	3	0	249.1	0.062	sheet flow
d1-8	0.3	0.55	3	0	301.4	0.062	sheet flow
d1-9	0.3	0.60	3	0	358.7	0.062	sheet flow
d1-10	0.3	0.65	3	0	421.0	0.062	sheet flow
d1-2uc05	0.3	0.20	3	0.05	42.3	0.062	ripple
d1-2uc10	0.3	0.20	3	0.10	49.8	0.062	ripple
d1-2uc15	0.3	0.20	3	0.15	62.3	0.062	ripple
d1-2uc20	0.3	0.20	3	0.20	80.0	0.062	ripple
d1-2uc25	0.3	0.20	3	0.25	102.1	0.062	ripple
d1-2uc30	0.3	0.20	3	0.30	129.5	0.062	ripple
d1-9uc05	0.3	0.60	3	0.05	361.2	0.062	sheet flow
d1-9uc10	0.3	0.60	3	0.10	368.7	0.062	sheet flow
d1-9uc15	0.3	0.60	3	0.15	381.1	0.062	sheet flow
d1-9uc20	0.3	0.60	3	0.20	398.6	0.062	sheet flow
d1-9uc30	0.3	0.60	3	0.30	448.4	0.062	sheet flow
d1-9uc40	0.3	0.60	3	0.40	518.2	0.062	sheet flow
d1-9uc50	0.3	0.60	3	0.50	607.8	0.062	sheet flow
d1-9uc60	0.3	0.60	3	0.60	717.4	0.062	sheet flow
d2-9	0.3	0.60	3	0	358.7	0.045	sheet flow

Note: the advection term is not included during simulation.

For the rippled bed cases, as the flow dynamics increase, the ripple height and ripple length increase first and cause the HCL to become thicker, then they decrease after reaching a maximum (case d1-3), see Figure 4-30. However, for case d1-3 to d1-6, although the ripple height decreases, the wave boundary thickness increases with stronger flow dynamics, and the reference concentration becomes larger, which induces a thicker HCL; however, suspended sediment concentration above the HCL is lower from case d1-3 to d1-6, which is caused by the lower ripple height. Compared case d1-6 (rippled bed) with case d1-8 (sheet flow), although the flow dynamic is stronger in case d1-8, the sediment concentration is lower, and the suspension height is lower than that for d1-6 because of the lower eddy viscosity (Figure 4-29) and thinner wave boundary layer (Figure 4-28). From d1-7 to d1-10, with stronger flow dynamics, the sediment concentration increases again, and the HCL develops when the BBL becomes larger too. Thus, it is concluded that we could directly establish the relation between the HCL and the BBL, as the BBL is affected by both bed forms and flow dynamics. It is not appropriate to relate the HCL with a single factor, e.g., simple ripple parameters or flow parameters.

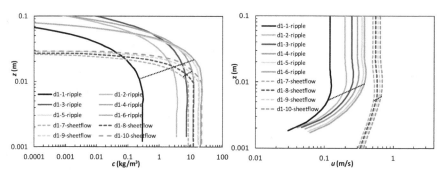

Figure 4-28. *Sediment concentration profiles (left) and maximum orbital velocity profiles (right) near the bed*

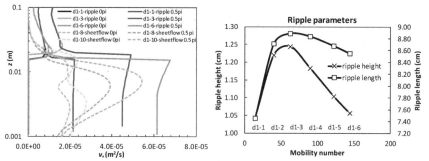

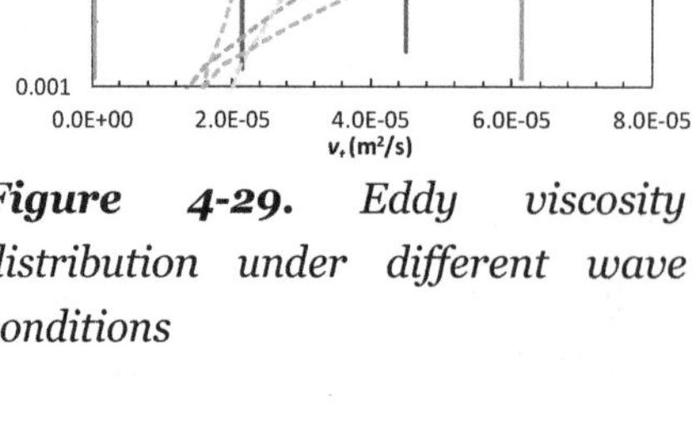

Figure 4-29. *Eddy viscosity distribution under different wave conditions*

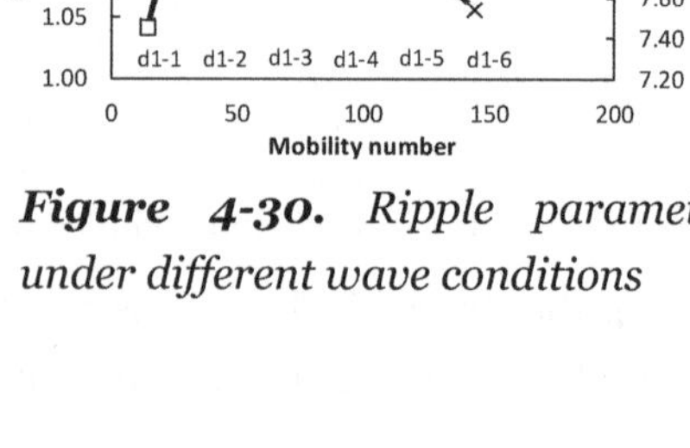

Figure 4-30. *Ripple parameters under different wave conditions*

(2) The role of waves and currents on the HCL

When a current is imposed, i.e., under combined wave-current conditions, the sediment concentration profile changes significantly, with much higher suspension and even an uniform value across the entire water depth, which has been proven by many experiments, for example, Yao et al. (2015) and Zhou and Ju (2007). Figure 4-31 shows the vertical distribution of the sediment concentration profile, eddy viscosity and maximum velocity under different combined wave-current conditions. The ratio of the current velocity to the wave max orbital velocity varies from 0 to 1. It can be seen that the sediment concentration near the bottom is dominated by waves and beyond the wave boundary layer the sediment concentration is greatly influenced by currents. The eddy viscosity plays a dominating role in sediment suspension. With different currents, the eddy viscosity changes little in the wave boundary layer, but changes significantly in the upper part, which corresponds to the changes of sediment concentration profile. It needs to be stated that the mobile bed effects were considered in the calculation, which may change the wave boundary layer under different wave-current conditions. See the velocity distributions in Figure 4-31.

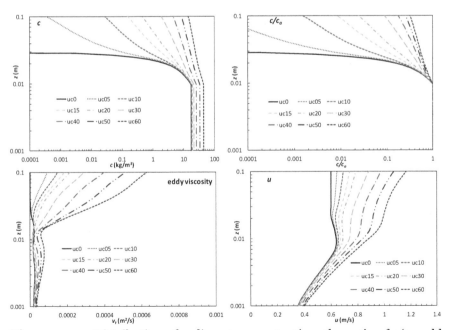

Figure 4-31. *Distribution of sediment concentration, the ratio of c/c$_a$, eddy viscosity and maximum velocity under different wave-current conditions under sheet flow conditions (for waves: u$_m$ = 0.6 m/s, T = 3 s, for currents:*

the depth-averaged velocity is represented by uc0 = 0.0 m/s, uc05 = 0.05 m/s, uc10 = 0.10 m/s, uc15 = 0.15 m/s, uc20 = 0.20 m/s, uc30 = 0.30 m/s, uc40 = 0.40 m/s, uc50 = 0.50 m/s, and uc60 = 0.60 m/s)

Over rippled beds, the changes of sediment concentration distribution have similar tendency with flat bed under different wave-current conditions. The evidence of experimental data has been shown by Yao et al. (2015) and Zhou and Ju (2007), and the data of Yao et al. (2015) is shown in Figure 4-32. To avoid the influence of ripples, sensitivity calculations were carried out by fixing the ripple parameters and changing only the current velocity. The results are shown in Figure 4-33. It can be seen that the changes of the sediment concentration profile are similar to those under sheet flow conditions. The vortex layer did not change as the ripples were fixed, thus the sediment concentration near the bottom did not change. Beyond the vortex layer, the sediment concentration increases significantly when currents are imposed.

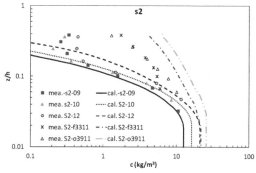

Figure 4-32. *Experimental data of sediment concentration profiles under waves and combined wave-current conditions (Yao et al. (2015), wave only cases: s2-09, s2-10, s2-12; wave-current cases: s2-f3311, s2-o3911)*

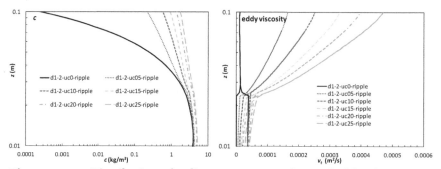

Figure 4-33. *Distribution of sediment concentration and eddy viscosity*

under different wave-current conditions over rippled bed (the ripple heights were fixed as the same)

4.4.2. Effects of bed forms on the SSC profiles

According to van Rijn (2007a), dune type bed forms are generally absent when the sediment bed is finer than about 100 µm, and the bed generally consists of a flat mobile surface or small-scale ripples. Baas et al. (2016) showed that ripples do exist for fine sediment or mixture sediment, not only for non-cohesive sediment but also for cohesive sediment. The dominant bed forms in oscillatory waves with or without a weak current in field conditions are often ripples with a length scale related (smaller or equal) to the near-bed orbital diameter. The impacts of bed forms on sediment transport have been studied by some scholars (Baas et al., 2016; Davies and Villaret, 2003; Grasmeijer and Kleinhans, 2004; Myrhaug and Holmedal, 2007; Styles and Glenn, 2003). In this paper we focus on the differences in simulation method and the shapes of concentration profile.

(1) Can we simulate the sediment concentration profile on rippled bed using the 'flat bed' method by adding the ripple-induced roughness?

To test the vortex's influence on the sediment concentration profile, Figure 4-34 shows the comparison of the sediment concentration profile calculated using the rippled bed method and the 'flat bed' method, as well as the experimental data for comparison. The flow dynamics conditions are the same in the simulation. When using the 'flat-bed' method, the ripple-induced roughness (varies from 22 mm to 29 mm in this case) was considered which was much higher than the grain roughness (only 0.155 mm). It can be seen that the flat bed method results in much lower concentration and fails to simulate the sediment concentration profiles. Under the same flow dynamics, the suspension on rippled beds by shedding of vortices is far greater than that of flat beds where no such coherent mechanism is present. The sediment concentration difference is several orders of magnitude between the two bed form conditions, especially in the upper part. This means that different approaches have to be employed to simulate the sediment concentration over different bed forms. The method which simply generalizes the ripples based on roughness cannot correctly simulate the sediment dynamics.

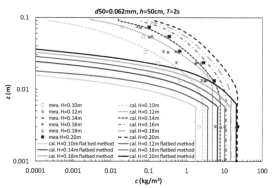

Figure 4-34. *Comparison of measured and calculated sediment concentration profiles under wave-only conditions (the measured date was Zhou and Ju (2007)'s case; the dot-lines were calculated using the 'rippled-bed' method, and the solid lines were calculated using the 'flat-bed' method)*

The experimental data of Yao et al. (2015) also showed this phenomenon. During the experiment of sediment s1 (d_{50} = 44 μm), there were ripples in wave-only cases, and the ripples disappeared when currents were imposed (case s1-f3212 and s1-03812). In this case, although the flow dynamics was stronger, the sediment concentration near the bottom decreased as the vortices disappeared (Figure 4-35). Since rippled beds occur in relatively low wave conditions, this can lead to the paradoxical outcome that, for a given mean current strength, more sediment may be transported in the presence of small waves above rippled beds than by sheet flow beneath large waves above plane beds (Davies and Thorne, 2005; Davies and Villaret, 2002).

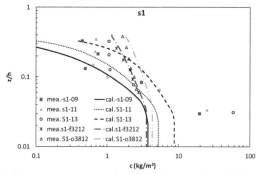

Figure 4-35. *Sediment concentration profiles under different conditions by Yao et al. (2015) (s1-09, s1-11, s1-12 were rippled bed cases. Currents were imposed in case of s1-f3212 and s1-03812)*

(2) The shape of sediment concentration profile over different bed forms

Are the shapes of the sediment concentration profiles similar over rippled bed and sheet flow conditions? To answer this question, Figure 4-36 shows the sediment concentration profiles over different bed forms. It can be concluded that the profile shapes are different with different bed forms. The shape of HCL is determined by bed forms, while the value is determined by flow dynamics. It is not difficult to explain this phenomenon mainly because of the eddy viscosity distribution (Fig. 4-29) which was discussed earlier.

Fig. 4-37 shows the time-series of sediment concentration contour over different bed forms. It can be seen that, under sheet flow conditions, the maximum concentration at the bottom happens nearly at the phase of maximum flow shear dynamics. While above rippled bed, it happens at the time of flow reversal because of the effects of the vortex. Besides, the figures also show the phase defection of sediment concentration in the upper part.

Bed forms are as important as flow dynamics for sediment transport, which means that we should not only analyze sediment transport with flow dynamics but also should consider bed forms. In short, bed forms determine the shape of concentration profile near the bottom, and flow dynamics determines the value of sediment concentration over a bed type.

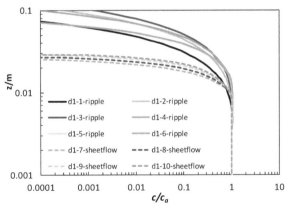

Figure 4-36. *Sediment concentration profiles over different bed forms (Lines represent rippled bed, and dash lines represent sheet flow)*

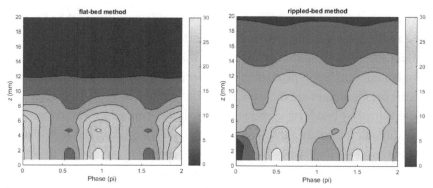

Figure 4-37. *Sediment concentration process within a wave period over different bed forms (unit: kg/m³, left: using flat-bed method, right: using rippled-bed method, H = 0.2 m, T = 2 s, and d_{50} = 0.062 mm)*

4.4.3. Stratification effects and mobile bed effects on sediment concentration profile

Stratification effects have been proved to have impacts on flow dynamics and sediment concentration profiles of mud (Winterwerp, 2001), and the results showed that stratification effects cause collapse of turbulence and greatly reduce the mud sediment concentration. From the experimental data of Dohmen-Janssen et al. (2001) and Zhou and Ju (2007), as well as assumed simulation cases, we did a preliminary study on the stratification effects. The sediment grain size varied from medium sand d_{50} = 0.21 mm, fine sand d_{50} = 0.13 mm to silt d_{50} = 0.062 mm. Figure 4-38 - Figure 4-40 show the comparison of the sediment concentration with and without stratification effects. It can be seen that stratification effects decrease the turbulence and concentration profile, which is in line with the general understanding. The finer sediment grain size has more significant stratification effects. Thus, stratification is a non-negligible factor for silt and very fine sand.

Figure 4-40 shows the comparison of the sediment concentration and eddy viscosity with and without stratification effects for silt of d_{50} = 0.062 mm and 0.045 mm. The results show that the eddy viscosity decreases greatly above the wave boundary layer, but changes little near the bottom. Above the HCL, the stratification decreases the eddy viscosity (or sediment diffusivity) greatly, and shows a collapsing behavior, which is similar to fluid mud (Winterwerp, 2001). The damping of turbulence contributes to the formation of the HCL, as the decreased diffusivity cannot sustain the sediment suspension. As a result, the concentration gradient becomes larger, which further increases the stratification effects. After equilibrium, a clear interface forms between the HCL and the upper clear water layer. However, near the bottom, the damping effects show little change, which shows the maintenance of the turbulence production. This is mainly because there is no flocculation process and the bottom consists of a consolidated layer, which is similar to sand but different from fluid mud. The comparison of d_{50} = 0.062 mm and 0.045 mm shows that finer sediment causes larger damping effects. Thus, the stratification behavior of silt has the transitional behavior between sand and cohesive mud, i.e., unlike sand, the stratification effects cannot be neglected; however, unlike fluid mud, the stratification effects for silt is not strong enough to destroy the flow dynamics.

The flux Richardson number Ri_f or bulk Richardson number Ri_c is often used to describe the stratification effects. Winterwerp (2001) argued that a turbulent shear flow collapses when the flux Richardson number exceeds a critical value which was found to be a constant (0.15) under steady state. In Yao et al. (2015)'s experiment, it was concluded that the critical value of the bulk Richardson number can be affected by the grain size and it is difficult to relate the Richardson number with silt-enriched concentration.

From Dohmen-Janssen et al. (2001)'s case and the assumed cases, we also tested the mobile bed effects on the sediment concentration profile in sheet flow conditions (see Figure 4-38 and Figure 4-41). It can be seen that the mobile bed affects the sediment concentration profile, i.e., taking the mobile bed roughness into consideration, the sediment concentration increases. In Figure 4-38, the grain roughness and mobile bed roughness are 0.32 mm and 1.48 mm, respectively, for D1-0.13 mm. The grain roughness and mobile bed roughness are 0.53 mm and 1.88 mm, respectively, for D2-0.21 mm. The ratio of mobile bed roughness to grain roughness is 4.6 for D1-0.13 mm and 3.5 for D2-0.21 mm. For finer sediment, for example, silt with d_{50} = 0.062 mm and d_{50} = 0.045 mm in Figure 4-41, the grain roughness is only 0.155 mm and 0.113 mm, respectively, which is too small and unrealistic. Therefore, the mobile bed roughness must be used in these cases, with the value of 0.58 mm and 0.47 mm, respectively. Because the grain roughness is too small to run the model, here a 2.5 times of grain roughness was used as the case of without mobile bed effects in Figure 4-41. Results show that the sediment concentration near the bottom without mobile bed effects was 21-29% smaller than that with mobile bed effects. Thus, for fine sediment, the mobile bed roughness is dominant.

The intensive sheet flow layer leads to turbulence damping and increased flow resistance. The damping of turbulence decreases the sediment concentration while the mobile bed effects increase the sediment concentration.

For medium sand, the effects of stratification and mobile bed may be neglected, and some models received good results without considering them (Holmedal et al., 2004; Zhang et al., 2011). However, for silt and fine sand, these physical processes are important impact factors, which cannot be neglected.

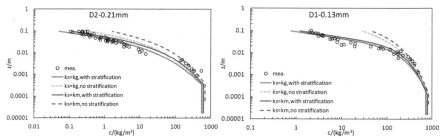

Figure 4-38. *Comparison of sediment concentration and velocity distribution: with or without stratification and with or without moving bed roughness (kg = grain roughness and km = mobile bed roughness. The experimental data came from Dohmen-Janssen et al. (2001))*

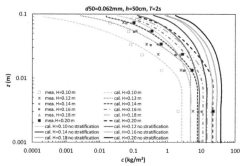

Figure 4-39. *Comparison of sediment concentration with and without stratification effects for silt with d_{50} = 0.062 mm (The measured data was from Zhou and Ju (2007)'s experiment over rippled-bed)*

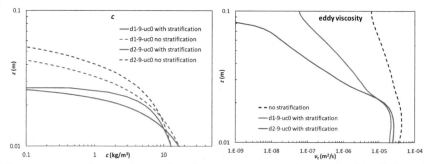

Figure 4-40. *Comparison of sediment concentration and eddy diffusivity with and without stratification effects (d_{50} = 0.062 mm (d1-9), d_{50} = 0.045*

mm (d2-9), T = 3 s, and u_m = 0.6 m/s)

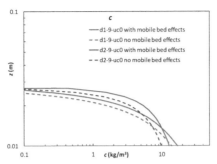

Figure 4-41. *Comparison of sediment concentration with and without mobile bed effects (d_{50} = 0.062 mm (d1-9), d_{50} = 0.045 mm (d2-9), T = 3 s, and u_m = 0.6 m/s)*

4.4.4. Discussion

For silt and fine sand, in wave-dominated environment, there is a high sediment concentration layer (HCL) near the bed bottom, which is about twice the height of the wave boundary layer. From above sensitivity analysis, we can see that, one of the main reasons of the formation of the HCL is that the eddy viscosity is limited within the BBL and it is difficult for the sediments to suspend to the upper part. The stratification effects contribute to the formation as the sediment diffusivity is further decreased from turbulence damping. Another reason lies in the hindered settling effects which induces lower effective settling velocity of the sediment lower in the water column than that in the upper part, similar to the so-called lutocline (Winterwerp, 1999).

Silt shows transition behavior between sand and cohesive clay. Winterwerp (2001) has elaborately described the behavior of non-cohesive and cohesive sediment. Suspensions of non-cohesive sediment under steady state conditions are characterized by equilibrium concentrations. The turbulence damping effect is not strong on deposited sand rigid bed, and turbulence production remains possible. Although there is still likely a very thin HCL on the sand bed, due to the relatively high settling velocity, the HCL cannot fully develop and bed load sediment transport is the main movement type. For cohesive sediment, because of flocculation processes, the deposited sediments do not form a rigid bed but cause a layer of fluid mud to form, thus create a two-layer fluid system. At the interface between the two-layer fluid, vertical turbulent mixing is strongly damped and results in a catastrophic collapse of the vertical turbulence field and the vertical sediment concentration profile.

Even under steady current or combined wave-current conditions, a distinct interface of fluid mud can still be investigated. Due to the strong damping effects, the HCL of silt may share a similar two-layer system as fluid mud. However, because of little flocculation of silt, although there is a turbulence collapse above the HCL, the turbulence in the wave boundary layer near the bottom can still be maintained. The equilibrium concentration concept may still be applicable for silt, but the stratification effects have to be included. Another important difference to fluid mud is that the HCL of silt only exists in wave dominant conditions.

4.5. Conclusion

A 1DV model was developed for flow dynamics and sediment transport in the wave-current bottom boundary layer, especially for simulation of the HCL of silt and very fine sand. Based on the physical background, special approaches for sediment movement were introduced, including approaches for different bed forms (rippled bed and 'flat-bed'), hindered settling, stratification effects, mobile bed effects, reference concentration and critical shear stress. For rippled beds, the combined vortex and k-ε model was employed to simulate the turbulence and the k and ε values at the interface of the vortex-dominated layer were derived. The approaches of hindered settling were employed considering the difference between silt and sand. An expression of silt-sand incipience of motion was employed for the critical shear stress. During the reference concentration calculation, it is unrealistic to give a zero value for the bed concentration during the stage of a wave cycle, when the shear stress is lower than the critical value. To overcome this shortcoming, the reference concentration was revised by considering the deposited sediment from the last time step. A number of experimental datasets were used to verify the model, which showed that the model is able to simulate the flow dynamics and sediment profiles reasonably, for sheet flow conditions and rippled bed, as well as silt and sand.

A HCL near the bottom is one of the most important characteristics of silty sediments. Sensitivity analysis was carried out on the factors impacting the HCL, i.e., bed forms, flow dynamics, and effects of stratification and mobile bed. The results showed that 1) the HCL is affected by both flow dynamics and bed forms and it is not appropriate to relate the HCL with a single factor; 2) as the BBL is the combined result of bed forms and flow dynamics, we could directly establish the relation between the HCL and the BBL; and 3) the thickness of the HCL is about twice the height of the wave boundary layer. Bed forms determine the shape of the concentration profile near the bottom, and flow dynamics determine the magnitude. Different approaches have to be employed to simulate the sediment concentration over different bed forms. The stratification effects impact the sediment concentration greatly and more works need to be carried out to investigate the stratification behavior of silt. For finer sediment, the mobile bed effects are larger because of a higher transport rate, i.e., the grain roughness is very low and the mobile bed roughness is dominant.

The simulation of the HCL helps us better understand the vertical concentration distributions of silt-dominated sediment under different wave-current conditions. It is a supplemental tool to flume experiments and a forerunner of 3D simulations. Meanwhile, it can serve as a simple reference model to test theoretical formulas, or to help assess the empirical parameterizations in those formulas for 2DH/3D modelling.

In natural environments, there are generally mixtures of clay, silt and sand. Bed composition will have effects on bed forms and sediment concentration distribution. At this stage, this paper mainly studies the high concentration behaviour of pure silt and very fine sandy sediments while the sediment composition was treated as uniform. It is a future study direction to simulate sediment mixtures.

Chapter 5

The mean SSC of silty sediments under wave-dominated conditions

The mean suspended sediment concentration (SSC) is one of the fundamental issues for engineering practice and numerical modelling. The parameterization of the SSC profile of silty sediment is still under-researched. This chapter focuses on the mean SSC profile and the depth-averaged concentration for silty sediment under non-breaking wave-dominated conditions. Firstly, the time-averaged diffusion equation for suspended sediment transport was analytically solved based on assumed distribution of sediment diffusivity, and the expressions for the mean SSC profile were derived. Secondly, the formula for the depth-averaged sediment concentration under wave conditions was yielded by integrating the SSC profile. The expression involves some basic physical processes, including the effects of bed forms, stratification and hindered settling. Verification using a number of experimental datasets showed that the proposed expressions can properly calculate the mean SSC for silt and are applicable for sand as well.

5.1. Introduction

For engineering practice or 2D/3D numerical simulation, it is imperative to know the expressions of the time-averaged SSC profile as well as the depth-averaged concentration. However, the parameterization for silty sediment remains understudied. Since the early 1900s, many scholars have studied the expressions for sand's SSC profile (Jayaratne et al., 2011; Liu, 2007; Nielsen, 1992; Nielsen, 1995; Rouse, 1937; Sleath, 1982; van Rijn, 2007a; Winyu and Shibayama, 1995; Zheng et al., 2013), but few investigated silty sediment. To develop a complete set of time-averaged SSC expressions has been a challenging task, due to the complexity of the suspension mechanism (Jayaratne et al., 2015).

The SSC profile is often described as an expression of a reference concentration and a shape function (Bolaños et al., 2012). The reference concentration, which has been studied extensively by many scholars, is specified close to the bed and provides the absolute level of the suspended load (Nielsen, 1992; van Rijn, 2007a; Zyserman and Fredsøe, 1994). The shape function represents the distribution profile with height above the bed and is normally derived from the sediment diffusivity distribution. Commonly, three kinds of sediment diffusivity distribution are known, (i.e., uniform, linear and parabolic), which induce different type of sediment concentration profile, (i.e., exponential, power and Rouse, respectively) (Soulsby, 1997). Many formulas for sediment concentration profile were proposed through assuming sediment diffusivity distribution (Coleman, 1969; Ravindra Jayaratne and Shibayama, 2007; Rouse, 1937; Umeyaina, 1992; van Rijn, 1993; Winyu and Shibayama, 1995). Lundgren (1972) proposed a toe-type distribution of wave-related viscosity. Nielsen (1992, 1995) argued that pure gradient diffusion was unsatisfactory and proposed a combined convection diffusion model by introducing a sediment mixing length and a convective function.

Previous studies on sand sediment concentration profile provide the methodology for further studying silt sediment. The equilibrium concept, for which instantaneous consolidation and relatively trivial turbulence damping effects were involved, is widely accepted for the sand regime. It is arguable whether the equilibrium concentration concept is suitable for silt sediment or not. For the cohesive mud sediment, the saturation concentration concept is used due to consolidation and flocculation process, but this is not the case for the equilibrium concept (Winterwerp and Van Kesteren, 2004). Mehta and Lee (1994) suggested that the 10-20 µm size may be considered practically to be the dividing size that differentiates cohesive and cohesionless sediment suspension behavior. Certain experiments (Li, 2014; Yao et al., 2015; Zhou and Ju, 2007) have shown that sediment with grain size of 45 μm to 110 μm shared similar suspension behavior under wave-current conditions. Therefore, it can be concluded that the suspended concentration profiles of medium silt and very fine sand can be treated similarly. The so-called 'equilibrium' concentration in silt regime cannot be established directly with flow conditions because of the flow-sediment interaction; however, from the results of chapter 4, a steady concentration could still be formed by introducing the stratification effects. In this case, involving silt, the concept of equilibrium concentration is employed.

This chapter aims to parameterize the sediment concentration profile as well as the depth-averaged concentration of silt and very fine sand sediments under non-breaking wave-dominated conditions. Firstly, distributions of wave-related sediment diffusivity over different bed forms were proposed inspired by the 1DV model in chapter 4; then, the time-averaged diffusion equation for suspended sediment was analytically solved by considering several important physical processes. This study is expected to assist in better understanding the SSC profile of silt and very fine sand as well as provide approaches for 2DH or 3D models.

5.2. Methods and Materials

5.2.1. Derivation method

The classic time-averaged governing equation for suspended sediment transport can be solved analytically to obtain the vertical distribution of SSC,

$$w_s \overline{c}(z) + \varepsilon_s \frac{d\overline{c}}{dz} = 0 \qquad (5\text{-}1)$$

in which w_s is settling velocity, ε_s is sediment diffusivity coefficient, $\overline{c}$ is mean sediment concentration and z is vertical coordinate.

The particular solution to Eq. (5-1):

$$\overline{c}(z) = \overline{c_a} e^{-G(z)} \quad \text{with} \quad G(z) = \int_{z_a}^{z} \frac{w_s}{\varepsilon_s(z)} dz \qquad (5\text{-}2)$$

in which, $\overline{c_a}$ is time-averaged reference concentration at reference height z_a. This solution depends on $\overline{c_a}$ and the distribution of sediment diffusivity. The formula for $\overline{c_a}$ of silt-sand is employed; it was proposed by van Rijn (2007a) and was extended to silt regime by Yao et al. (2015). Appendix C illustrates the details of the reference concentration.

(1) Sediment diffusivity

Under combined wave-current conditions, the combined sediment diffusivity is given by the square sum of the wave-related and current-related diffusivities (Nielsen, 1992; van Rijn, 2007a).

$$\varepsilon_{s,cw} = \sqrt{\varepsilon_{s,c}^{2} + \varepsilon_{s,w}^{2}} \qquad (5\text{-}3)$$

in which $\varepsilon_{s,cw}$ = combined sediment diffusivity coefficient, $\varepsilon_{s,w}$ = wave-related sediment diffusivity coefficient, and $\varepsilon_{s,c}$ = current-related sediment diffusivity coefficient.

van Rijn (2007a)'s formula is employed for the current-related sediment diffusivity,

$$\varepsilon_{s,c} = \begin{cases} \kappa\beta_{w}u_{*_c}z(1-z/h) & z < 0.5h \\ 0.25\kappa\beta_{w}u_{*_c}h & z > 0.5h \end{cases} \qquad (5\text{-}4)$$

in which u_{*_c} = current-related shear velocity, $\beta_w = \max[1.5, 1 + 2(w_s / u_{*_c})^2]$.

The distribution of wave-related sediment diffusivity/eddy viscosity is proposed over different bed forms, inspired by the 1DV model for wave-current bottom boundary layer in chapter 4. Details will be presented in the following sections.

(2) Stratification effects

In turbulence models, the buoyancy flux B_k can be introduced to simulate the stratification effects. However, to derive a parameterized expression, the stratification effects are considered by introducing the turbulence damping coefficient, $\varepsilon_{sm}(z) = \phi_d \varepsilon_s(z)$ (van Rijn, 2007a).

$$\phi_d = \phi_{fs}[1 + (c_v / c_{gel,s})^{0.8} - 2(c_v / c_{gel,s})^{0.4}] \qquad (5\text{-}5)$$

with $\phi_{fs} = d_{50} / (1.5d_{sand})$, and $\phi_{fs} = 1$ for $d_{50} \geq 1.5d_{sand}$. c_v is volume sediment concentration of solids, $c_{gel,s}$ = 0.65 = maximum bed concentration, and d_{sand} = 0.062 mm.

(3) Hindered settling

For sand, according to Richardson and Zaki (1954) and van Rijn (2007a), the settling velocity in a fluid-sediment suspension can be determined as:

$$w_s = w_{s,0}(1 - c_v)^n \qquad (5\text{-}6)$$

For silt and very fine sand (Te Slaa et al., 2015):

$$w_s = w_{s,0} \frac{(1 - c_v / \phi_{s,struct})^m (1 - c_v)}{(1 - c_v / \phi_{s,max})^{-2.5\phi_{s,max}}} \qquad (5\text{-}7)$$

in which $w_{s,0}$ is settling velocity in clear water and the formula of van Rijn (2007a) was employed; n is the exponent, varying from 4.6 to 2.3; $\phi_{s,struct}$ = 0.5

is the structural density; $\phi_{s,\max} = 0.65$ is the maximum density, and $m = 1\text{-}2$.

Then, the solution of Eq. (5-1) turns to:

$$\bar{c}(z) = \bar{c}_a e^{-G(z)} \quad \text{with} \quad G(z) = \int_{z_a}^{z} \frac{w_s}{\phi_d \varepsilon_s(z)} dz \tag{5-8}$$

(4) Depth-averaged sediment concentration

The depth-averaged sediment concentration is another important parameter, which is frequently used in engineering practice and in the 2DH sediment simulation. The equilibrium depth-averaged sediment concentration is also referred to carrying capacity of the flow-dynamics. Some formulas were proposed by different approaches, such as energy balance theory (Dou et al., 1995; Xia et al., 2011; Zhang et al., 2009) or a synthetical method of dimensional analysis and measured data fitting (Liu, 2009). Given the vertical distribution of concentration, the equilibrium sediment concentration can be obtained by integrating the sediment flux or concentration profile in vertical (van Rijn, 1984b). The way of integrating sediment profile in vertical has clear physical significance, as it fully considers the mechanism of sediment exchange near the bed and distribution in the water volume.

The mean depth-averaged sediment concentration $\bar{c}_h$ follows from integrating the sediment concentration profile over water depth.

$$\bar{c}_h = \frac{1}{h - z_a} \int_{z_a}^{h} \bar{c}(z) dz \tag{5-9}$$

5.2.2. Materials

Experimental data of Horikawa et al. (1982), Havinga (1992), van Rijn et al. (1993), Ribberink and Al-Salem (1995), Williams et al. (1998), Dohmen-Janssen et al. (2001), O'Donoghue and Wright (2004), Zhou and Ju (2007), Li (2014), and Yao et al. (2015) were used to study and verify the proposed expressions, as listed in Table 5-1. The bed forms included 'flat bed' (sheet flow) and rippled bed; the flow dynamics included wave only cases and combined wave-current cases; and the sediment materials included silt and sand. The field datasets in Caofeidian sea area (Zuo et al., 2014) and Huanghua port sea area (Zhao and Han, 2007), described in chapter 2, were used for evaluation, where the sediment concentrations were observed during wave events in silt-dominated sea areas.

Table 5-1. *Collected data for sediment concentration*

Source	Flow dynamics	Data type	d_{50} (mm)	Bed forms
Horikawa et al. (1982)	Wave	Oscillatory tunnel	0.20	Flat bed
Havinga (1992)	Wave+current	Wave flume	0.10	Rippled bed
van Rijn et al. (1993)	Wave+current	Wave flume	0.11-0.22	Rippled bed
Ribberink and Al-Salem (1995)	Wave	Oscillatory tunnel	0.21	Flat bed
Willams et al. (1998)	Wave	Wave flume	0.329	Rippled bed
Dohmen-Janssen et al. (2001)	Wave+current	Wave flume	0.13-0.32	Flat bed
O'Donoghue and Wright (2004)	Wave	Oscillatory tunnel	0.15-0.28	Flat bed
Zhou and Ju (2007)	Wave	Wave flume	0.062-0.11	Rippled bed
Li (2014)	Wave	Wave flume	0.045-0.11	Rippled bed
Yao et al. (2015)	Wave and Wave+current	Wave flume	0.046-0.088	Rippled bed & flat bed
Zuo et al. (2014)	Wave+current	Field data	0.015	Flat bed
Zhao and Han (2007)	Wave+current	Field data	0.036	Flat bed

5.3. Mean sediment concentration profile over different bed forms

5.3.1. Time-averaged SSC profile over flat bed

5.3.1.1. Mean wave-related sediment diffusivity over flat bed

Many turbulence models can well simulate the eddy viscosity instantaneously, such as the k-ε model. However, for analytical analysis, the time-averaged distribution of eddy viscosity is expected to be known. The distribution of the time-averaged wave-related eddy viscosity was proposed, inspired by the intra-wave 1DV model in chapter 4, which solved the k-ε turbulence model. In the non-breaking wave conditions, a three-layer distribution is proposed, (i.e., linear at lower BBL, parabolic at middle part and uniform at upper part):

$$\overline{\upsilon_{t,w}}(z) = \begin{cases} \kappa u_*'z & z \le 0.5\delta_w \\ \kappa u_*'z(a_f - b_f z/2.5\delta_w) & 0.5\delta_w < z < 2.5\delta_w \\ r_s \kappa u_*'\delta_w & z \ge 2.5\delta_w \end{cases} \tag{5-10}$$

in which, $u_*' = 0.5u_{*w}$ = effective mean wave shear velocity, u_{*w} = wave maximum shear velocity, $\delta_w = \kappa u_* / \omega$ = thickness of wave boundary layer (Grant and Madsen, 1986), ω = wave frequency, $r_s = 0.8$, $a_f = 1.17$, $b_f = 0.85$. Then the wave-related sediment diffusivity $\overline{\varepsilon_{s,w}} = \overline{\upsilon_{t,w}}/\sigma$, with $\sigma = 1$.

Figure 5-1 shows the comparison of the eddy viscosity computed by Eq. (5-10) with the value from the 1DV model. It can be seen that, the proposed distribution has similar tendency with the results of the 1DV model. The parameterized distribution fits the averaged value, though there are deviations in the upper part for different cases.

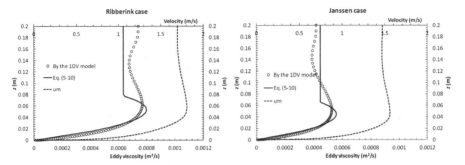

Figure 5-1. *Comparison of the distribution of the eddy viscosity coefficient computed by Eq. (5-10) with the results from the 1DV model (the calculated wave conditions were after Ribberink and Al-Salem (1995) and Dohmen-Janssen et al. (2001); the results are under wave only conditions without effects of sediment)*

5.3.1.2. The sediment concentration profile over flat bed

Substituting the expression of wave-related sediment diffusivity to Eq. (5-8), then, yields the distribution of sediment concentration under wave conditions,

$$
\overline{c}(z) = \begin{cases} \overline{c_a}(\dfrac{z}{z_a})^{-\alpha_f} & z_a < z \le z_1 \\[2ex] \overline{c_{z1}}(\dfrac{z-h'}{z}\dfrac{z_1}{z_1-h'})^{\frac{\alpha_f}{\alpha_f}} & z_1 < z < z_2 \\[2ex] \overline{c_{z2}}\exp[-\dfrac{\alpha_f}{r_s\delta_w}(z-z_2)] & z \ge z_2 \end{cases} \qquad (5\text{-}11)
$$

in which,

$\alpha_f = \dfrac{w_s}{\phi_d \kappa u_*}$ = Rouse number over plan bed considering stratification effects,

$z_1 = \max(z_a, 0.5\delta_w)$, $z_2 = \max(z_a, 2.5\delta_w)$,

$\overline{c_{z1}} = \overline{c_a}(\dfrac{z_1}{z_a})^{-\alpha_f}$, $\overline{c_{z2}} = \overline{c_{z1}}(\dfrac{z_2-h'}{z_2}\dfrac{z_1}{z_1-h'})^{\alpha_f}$, and $h' = a z_2 / b$.

It can be seen that the distribution of SSC is a power law in the low part and an exponential law in the upper part.

Under combined wave-current conditions, the time-averaged SSC profile was calculated by numerical procedure,

$$\frac{d\bar{c}(z)}{dz} = -\frac{w_s \bar{c}(z)}{\phi_d \varepsilon_s(z)} \qquad (5\text{-}12)$$

For fine sediment, the grain roughness is very small and the roughness enhanced by mobile bed effects is dominated. Camenen et al. (2009)'s formula was employed for the enhanced roughness.

5.3.1.3. Verification

(1) Experimental cases

The experimental datasets (Dohmen-Janssen et al., 2001; Ribberink and Al-Salem, 1995; Yao et al., 2015) were collected to verify the SSC profile in 'flat bed' conditions. Actually, existing experimental datasets for 'flat bed' (sheet flow) are mainly for sand. One case for silt is Yao et al. (2015)'s experiment, i.e., the case of s1-f3212 and s1-03812 with d_{50} = 46 μm in combined wave-current conditions, the rippled bed was washed away. Figure 5-2 to Figure 5-4 show the verification of the experimental data, and the calculated value fit the measured data reasonably.

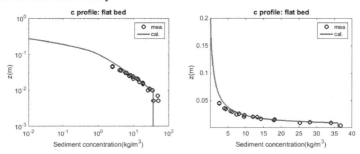

Figure 5-2. *Verification of the experimental data of Ribberink and Al-Salem (1995) (left: in log axis; right: in normal axis)*

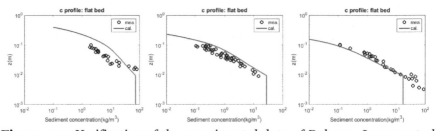

Figure 5-3. *Verification of the experimental data of Dohmen-Janssen et al. (2001) (left: (a) d_{50} = 0.13 mm, u_c = 0.24 m/s; middle: (b) d_{50} = 0.21 mm, u_c = 0.23m/s; right: (c) d_{50} = 0.32 mm, u_c = 0.26 m/s)*

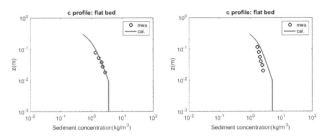

Figure 5-4. *Verification of the experimental data of Yao et al. (2015) (left: (a) case of s1-f3212, d_{50} = 0.046 mm; right: (b) case of s1-03812, d_{50} = 0.046 mm)*

(2) Evaluation on field data

(i) Caofeidian sea area

Figure 5-5 shows the comparison of the calculated and measured concentration at 0.4 m above the bottom in Caofeidian sea area during several wave events. The sediment median size is about 0.01~0.02 mm with average of 0.015 mm on the measurement site. Under calm conditions (with wave height < 0.5 m), sediment cannot be stirred up and sediment concentration was very low with the averaged value of only 0.05 kg/m3, which was assumed to be background concentration. The sediment concentration increased during windy days. It was assumed that the increase of sediment concentration may happen in some suitable conditions, e.g., when water depth was shallower and current velocity was larger. The calculated conditions were chosen as the shallowest water depth (2 m) and maximum flow velocity (0.57 m/s) on this site, with the wave height of 0.60-0.76 m.

It can be seen that, though there is considerable discrepancy between the calculated and measured SSC peaks in Figure 5-5, the magnitude is similar. The reason of the discrepancy many be as follows. First, this study is only for equilibrium concentration, while in the field, it is possible that sediment is suspended elsewhere (e.g., the shoal) and transported to this site, which is non-equilibrium. Second, the wave height was derived from wind speed by empirical methods, which might have caused the mismatch in phasing between waves (estimated) and concentrations (measured). Third, the sediment grain size Caofeidian is finer than the above experimental cases; however, silt is still the dominate part. We still need more data sets to verify finer sediment like in Caofeidian. Thus, we use "evaluation" instead of "verification", and only compare the order of magnitude of the SSC. Though there is large discrepancy in phase between the calculated and measured SSC, the magnitude is similar, around 0.1-0.3 kg/m3.

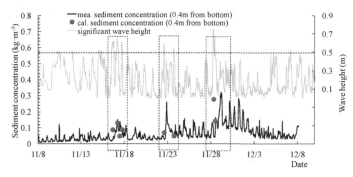

Figure 5-5. *Comparison of the calculated and measured concentration in Caofeidian sea area*

(ii) Huanghua port sea area

According to the measured data in 2003 in Huanghua port sea area (Zhan and Han, 2007), see section 2.3 in chapter 2, we evaluate the performance of the parameterized expression of Eq. (5-11). The "evaluation" is used here too, because of the missing of some details in the measured data, such as the process of the tidal level and the current velocities. The mean water depth is used as 6.4 m according to the bathymetry and mean tidal level. The medium size of bed material is 0.036 mm. Measurements show that high SSC occurs during storm surges, which causes heavy sudden siltation in navigation channels. The average tidal current without winds is about 0.4 m/s, but the current velocities during windy days are not found. According to the wind speed, we estimated the mean wave-driven current velocity. The wind shear stress $\tau_s = \rho_a C_d U_{wind}^2$, where ρ_a is air density, C_d is drag coefficient, U_{wind} is wind velocity. The mean measured wind velocity during that event is about 14 kg/m³. The calculated τ_s is about 0.38 N/m². From $\bar{u} = \sqrt{\tau_s C^2/(\rho g)}$, in which C is Chezy coefficient, the mean wind-induced current velocity can be estimated as 0.4 m/s. The total mean velocity is estimated as sum of absolute value of the mean wind-induced current velocity and the mean tidal current velocity, 0.8 m/s.

Figure 5-6 shows the comparison of the calculated and measured SSC profiles in Huanghua port sea area during windy days. It can be seen that, during November 5 to November 6, 2003, the agreement between the calculated and measured concentration is quite good. After November 7, the wind speed as well as wave height became smaller; however, the SSC can still remain a certain value, because it needs time for sediment to settle down and the concentration is over-saturated (Figure 5-6 (f), (g) and (h)). As the calculated value is an equilibrium one that could be seen as sediment capacity, it is reasonable that the calculated value is larger than the measured one. It is one of the causes of the heavy deposition in navigation channel, i.e., the SSC is much higher than the sediment capacity after a wind, and sediment settles in the channels where the flow dynamics are normally weak.

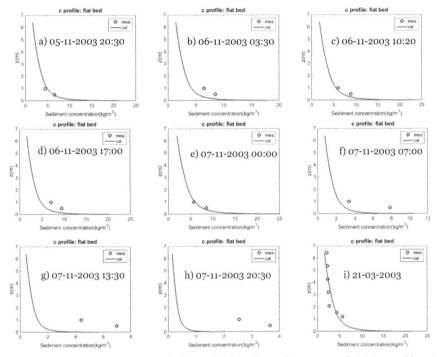

Figure 5-6. *Comparison of the calculated and measured SSC profiles in Huanghua port sea area during windy days in 2003*

5.3.2. Time-averaged sediment concentration profile over rippled bed

5.3.2.1. Mean wave-related sediment diffusivity over rippled bed

Over rippled bed, momentum transfer and the associated sediment dynamics in the near-bed layer are dominated by coherent motions, in particular the process of vortex formation above the ripple lee slopes and the shedding of these vortices at times of flow reversal (van der A, 2005). In a near-bed layer, approximately two ripple heights below, the flow dynamics are dominated by the periodic vortex structures, whereas above this layer the coherent motions break down and are replaced by random turbulence (Davies and Villaret, 1999). This leads to considerably higher height of sediment suspension compared to flat beds. According to this physical background, a two-layer model was adopted, see Figure 5-7, i.e., the vortex-dominated layer at the bottom and the turbulence-dominated layer above, separated by twice the ripple height (Davies and Thorne, 2005; van der Werf et al., 2006).

In the vortex layer ($z<2\eta$), Nielsen (1992)'s formula was employed for vortex, as referred by Davies and Thorne (2005),

$$\overline{\upsilon_{t,w}}(z) = \overline{\upsilon_{tN}} = c_v A\omega k_s \qquad\qquad z<2\eta \qquad\qquad (5\text{-}14)$$

in which, $\overline{\upsilon_{tN}} = c_{vor} A\omega k_s$ with c_{vor} = 0.004-0.005.

Above the vortex layer ($z>2\eta$), turbulence is dominated. Similar with the flat bed, the mean eddy viscosity was proposed to be a three-layer distribution (Eq. 5-15), with linear distribution at bottom, parabolic distribution in the middle part and uniform in the upper part. According to the comparison with the results of the 1DV model, Figure 5-8 shows that the proposed distribution is reasonable.

$$\overline{\upsilon_{t,w}}(z) = \begin{cases} \kappa u_{y*}z & 2\eta < z \leq 2.5\eta \\ \kappa u_{y*}z[a_r - b_r z / 4.5\eta] & 2.5\eta < z < 4.5\eta \\ b_{up}\kappa u_{y*}\eta & z \geq 4.5\eta \end{cases} \qquad (5\text{-}15)$$

in which $u_{y*} = \overline{\upsilon_{tN}} / \kappa / 2\eta$, b_{up} = 2.25, a_r = 1.625, b_r = 1.125.

The sediment diffusivity in the lower layer above rippled beds is significantly larger than the eddy viscosity, with $\overline{\varepsilon_{s,w}} = \beta\overline{\upsilon_{t,w}} / \sigma$ (Nielsen, 1992; Thorne et al., 2002). The coefficient β is given by

$$\beta = \begin{cases} 4 & z \leq 2\eta \\ 4 - 3(\dfrac{z-2\eta}{h-2\eta})^\gamma & z > 2\eta \end{cases} \qquad\qquad (5\text{-}16)$$

with the coefficient γ = 0.4-1.

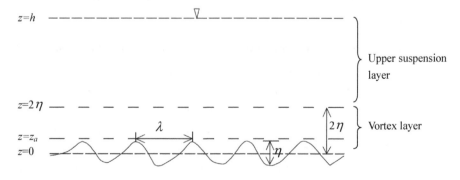

Figure 5-7. *The physical concept of a two-layer model over rippled bed*

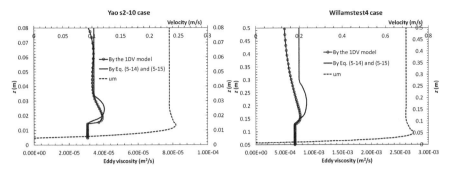

Figure 5-8. *Comparison of the distribution of the eddy viscosity calculated by Eq. (5-14) and Eq. (5-15) with the results of the 1DV model*

5.3.2.2. The SSC profile over rippled bed

Substituting the expressions for the wave-related sediment diffusivity over rippled bed to Eq. (5-8), then, yields the distribution of SSC under wave conditions. However, the expression is too complex to integrate when $\varepsilon_s(z) = \beta(z)\upsilon_t(z)/\sigma$ at $z > 2\eta$ is applied. An average value $\overline{\beta}$ was used in the domain z_{1r} to z_{3r} from the perspective of practice,

$$\overline{\beta_i} = \frac{1}{\Delta z}\int_{z_i}^{z_{i+1}}\beta(z)dz = 4 - \alpha_\beta[(z_{i+1}-2\eta)^{\gamma+1}-(z_i-2\eta)^{\gamma+1}]\frac{1}{\Delta z} \qquad (5\text{-}18)$$

with $\alpha_\beta = \dfrac{1}{\gamma+1}\dfrac{3}{(h-2\eta)^\gamma}$ and $\Delta z = z_{i+1}-z_i$.

The final expression for sediment concentration profile over rippled bed under wave conditions:

$$\overline{c}(z) = \begin{cases} \overline{c_a}\exp[-\dfrac{w_s}{4\phi_d\upsilon_{tN}}(z-z_a)] & z_a < z \leq z_{1r} \\[2mm] \overline{c_{z1r}}(\dfrac{z}{z_{1r}})^{-\alpha_r} & z_{1r} < z \leq z_{2r} \\[2mm] \overline{c_{z2r}}(\dfrac{z-h'}{z}\dfrac{z_{2r}}{z_{2r}-h'})^{\alpha_r/a_r} & z_{2r} < z < z_{3r} \\[2mm] \overline{c_{z3r}}d(z)^{\alpha_d} & z \geq z_{3r} \end{cases} \qquad (5\text{-}19)$$

in which,

$\alpha_r = \dfrac{w_s}{\phi_d\overline{\beta}\kappa u_{v*}}$ = Rouse number over rippled bed considering stratification

effects, $h' = a_r z_{3r}/b_r$, $z_{1r} = \max(z_a, 2\eta)$, $z_{2r} = \max(z_a, 2.5\eta)$, $z_{3r} = \max(z_a, 4.5\eta)$,

123

$$\overline{c_{z1r}} = \overline{c_a} \exp[-\frac{w_s}{\phi_d \varepsilon_s}(z_{1r} - z_a)] \ , \quad \overline{c_{z2r}} = \overline{c_{z1r}}(\frac{z_{2r}}{z_{1r}})^{-\alpha_r} \quad \text{and} \quad \overline{c_{z3r}} = \overline{c_{z2r}}(\frac{z_{3r} - h'}{z_{3r}} \frac{z_{2r}}{z_{2r} - h'})^{\frac{\alpha_r}{\alpha_r}}.$$

$$d(z) = \frac{4(h - 2\eta) - 3(z - 2\eta)}{4(h - 2\eta) - 3(z_{3r} - 2\eta)} \quad \text{and} \quad \alpha_d = \frac{w_s}{\phi_d \kappa u_{v*}} \frac{1}{b_{up}\eta} \frac{h - 2\eta}{3}.$$

It can be seen that the expression shows an exponential distribution in the vortex-dominated layer near the bottom and power distributions at the upper part. The expression has similar structure with Bolaños et al. (2012), who collected many experimental datasets and proposed the sand SSC profile formula by data fitting. Iteration is needed when using this equation as the stratification effects and hindered settling are included. Fortunately, the iteration was only about 5-7 times according to the verification cases. Under combined wave-current conditions, the approach is the same for flat bed, i.e., by solving the numerical procedure of Eq. (5-12).

5.3.2.3. Verification

Some experimental datasets were collected to verify the mean SSC profile over rippled bed, see Table 5-1, i.e., Havinga (1992), van Rijn et al. (1993), Zhou and Ju (2007), Li (2014) and Yao et al., (2015). These cases include sediment range of silt and sand, wave-only conditions and combined wave-current conditions. Figure 5-9 to Figure 5-13 show the calibration results. It can be seen that, though there were some deviations in some cases, the calculated value is in agreement with the measured data.

The measured SSC profile can be considered as a fully developed equilibrium profile under wave-only conditions, because of the relatively small net current. However, sufficient sediment source and a certain distance are needed to establish the equilibrium concentration when a current is added. The length of sediment section in flume experiments was normally not long enough to achieve equilibrium concentration (Yao et al., 2015). For example, the length of the sediment bed was 15 m and 25 m in Yao et al. (2015)'s experiment and van Rijn et al. (1993)'s experiment, respectively, and it was still too short to develop equilibrium concentration when relatively strong currents were imposed.

Under wave-only conditions, the calculated sediment concentration agrees well with the measured data; while under combined wave-current conditions, the calculated sediment concentration (equilibrium) is much larger than the measured value (non-equilibrium) under stronger current conditions. The non-equilibrium concentration may be further simulated by a 2DV model considering longitudinal diffusive transport. However, despite of the discrepancies between the computed and measured data, the proposed equations are able to simulate a straighter SSC profile as the current velocity increases.

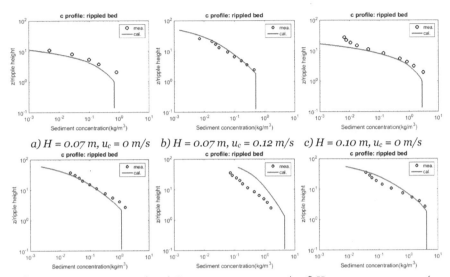

a) $H = 0.07$ m, $u_c = 0$ m/s   b) $H = 0.07$ m, $u_c = 0.12$ m/s   c) $H = 0.10$ m, $u_c = 0$ m/s

d) $H = 0.10$ m, $u_c = 0.12$ m/s   e) $H = 0.10$ m, $u_c = 0.24$ m/s   f) $H = 0.14$ m, $u_c = 0.12$ m/s

Figure 5-9. Verification of the experimental data of Havinga (1992) (h = 0.4 m, d_{50} = 100 μm)

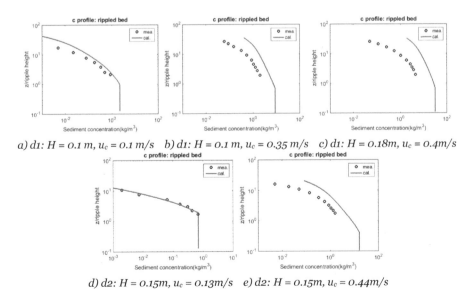

a) d1: $H = 0.1$ m, $u_c = 0.1$ m/s   b) d1: $H = 0.1$ m, $u_c = 0.35$ m/s   c) d1: $H = 0.18$m, $u_c = 0.4$m/s

d) d2: $H = 0.15$m, $u_c = 0.13$m/s   e) d2: $H = 0.15$m, $u_c = 0.44$m/s

Figure 5-10. Verification of the experimental data of van Rijn et al. (1993) (h = 0.4 m, d1: d_{50} = 110 μm and d2: d_{50} = 200-220 μm)

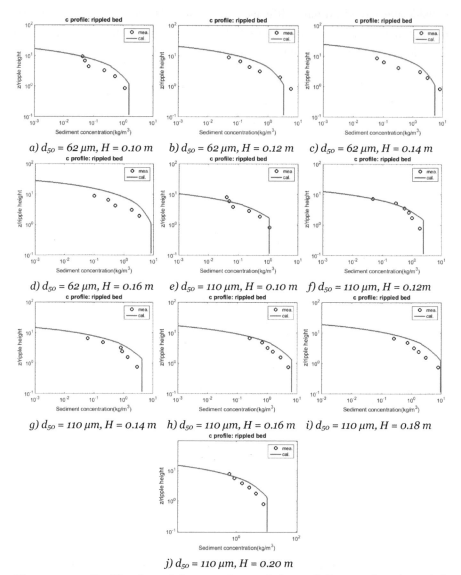

a) d_{50} = 62 μm, H = 0.10 m   b) d_{50} = 62 μm, H = 0.12 m   c) d_{50} = 62 μm, H = 0.14 m

d) d_{50} = 62 μm, H = 0.16 m   e) d_{50} = 110 μm, H = 0.10 m   f) d_{50} = 110 μm, H = 0.12m

g) d_{50} = 110 μm, H = 0.14 m   h) d_{50} = 110 μm, H = 0.16 m   i) d_{50} = 110 μm, H = 0.18 m

j) d_{50} = 110 μm, H = 0.20 m

Figure 5-11. *Verification of the experimental data of Zhou and Ju (2007) (h = 0.5 m, T = 2 s, u_c = 0 m/s)*

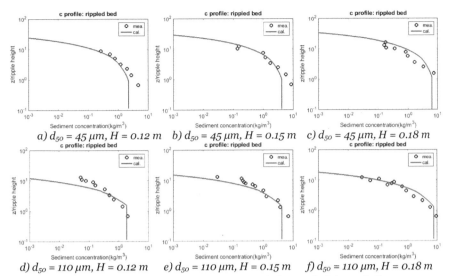

Figure 5-12. Verification of the experimental data of Li (2014) (h = 0.5 m, T = 2 s, u_c = 0 m/s)

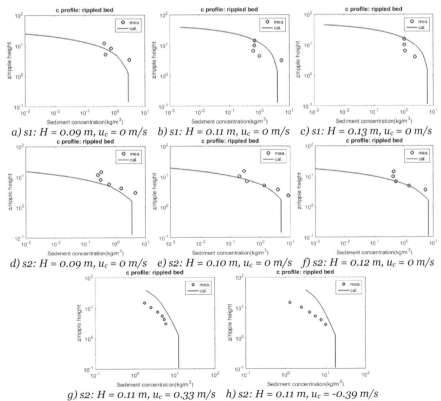

Figure 5-13. Verification of the experimental data of Yao et al. (2015) (h = 0.3 m, T = 1.5 s, s1: d_50 = 44 μm, s2: d_50=88 μm)

5.3.3. Discussion

5.3.3.1. The SSC profile over rippled but hydrodynamically plane bed

The above results over rippled bed are only valid for vortex ripples, with ripple steepness larger than 0.12. For the lower ripple steepness, the bed form is hydrodynamically plane (Davies and Thorne, 2005; van der Werf et al., 2006). This kind of rippled bed is treated as 'flat bed'. However, the roughness height is still calculated by the rippled bed method.

Figure 5-14 shows verification of some experimental cases of Zhou and Ju (2007), Havinga (1992) and van Rijn et al. (1993) with low ripple steepness and Table 5-2 shows the experimental conditions. It can be seen that the calculated sediment concentration profiles fit the measured data reasonably. The results indicate that the approaches can roughly represent the main physical background of this bed form. As this kind of bed type is a transition zone between vortex ripples and sheet flow, the turbulence diffusion is very complex and still needs further study.

Table 5-2. *Some experimental cases over bed forms with low ripple steepness*

Case	d_{50} (µm)	u_m (m/s)	u_c (m/s)	Mobility number	Ripple height(cm)	Ripple length cm)	Ripple steepness
Zhou and Ju (2007)	62	0.35-0.39	0.0	124-154	0.58-0.73	6.38	0.09-0.115
Havinga (1992)	100	0.31	0.240	94.6	0.79	7.45	0.106
van Rijn et al. (1993)	111	0.35	0.131	76.8	0.70	6.80	0.103

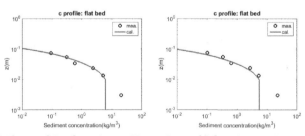

a) Zhou and Ju: d_{50} = 62 µm, H = 0.18 m   b) d_{50} =62 µm, H = 0.20 m

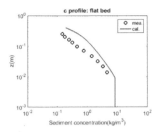 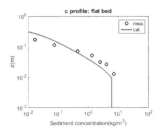

c) Havinga: H = 0.14 m, u_c = 0.24 m/s   d) van Rijn et al.: H = 0.18 m, u_c = 0.13 m/s

Figure 5-14. *Verification of the experimental data of Zhou and Ju (2007), Havinga (1992) and van Rijn et al. (1993)*

5.3.3.2. Comparison with other formulas

Two formulas for SSC profile were chosen to compare, the formula of van Rijn (2007a) and Nielsen (1992). van Rijn (2007a) proposed a distribution of sediment diffusivity, Eq. (5-20). The sediment concentration profile was derived from Eq. (5-20), considering the stratification effects and hindered settling.

$$\varepsilon_{s,w} = \begin{cases} \varepsilon_{s,bed} & z \leq 2\eta \\ \varepsilon_{s,vortex} + (\varepsilon_{s,w,max} - \varepsilon_{s,vortex})(\dfrac{z - \delta_s}{0.5h - \delta_s}) & 2\eta < z < 0.5h \\ \varepsilon_{s,w,max} & z \geq 0.5h \end{cases} \qquad (5\text{-}20)$$

in which, $\varepsilon_{s,w,bed} = 0.018\gamma_w\beta_w\delta_s U_{\delta,r}$ = wave-related sediment mixing coefficient near the bed, $U_{\delta,r}$ = representative near-bed peak orbital velocity based on significant wave height, $\beta_w = 1 + 2(w_s/u_{*,w})^2$ with $\beta_w \leq 1.5$; and $u_{*,w}$ = wave-related bed-shear velocity, $\gamma_w = 1 + (H_s/h - 0.4)^{0.5}$ = empirical coefficient related to wave breaking ($\gamma_w = 1$ when $H_s/h < 0.4$). $\varepsilon_{s,w,max} = 0.035\gamma_w hH_s/T_p$ with $\varepsilon_{s,w,max} \leq 0.05 m^2/s$.

The thickness of effective near-bed sediment mixing layer $\delta_s = 2\gamma_w\delta_w$ with limits $0.1 \leq \delta_s \leq 0.5m$. $\delta_w = 0.36A_\delta(A_\delta/k_{s,w,r})^{-0.25}$ = thickness of wave boundary layer, A_δ = peak orbital excursion based on significant wave height; and $k_{s,w,r}$ = wave-related bed roughness.

Nielsen (1992) proposed a formula for SSC profile, Eq. (5-21), considering advection effects of ripples,

$$\overline{c}(z) = \overline{c_a} \exp[-\frac{1}{L_s}(z - z_a)] \qquad (5\text{-}21)$$

in which, L_s is the vertical scale of the convective mixing process.

Figure 5-15 shows comparison of the SSC profiles using different formulas. It can be seen that, all formulas could simulate the sediment concentration profile well in the sand regime; however, in the silt range, the distribution of sediment diffusivity of van Rijn (2007) over-estimated while Nielsen (1992) low-estimated the SSC profile. The formulas of van Rijn (2007) and Nielsen (1992) were derived for sand regime and worked well in their application scope. Actually, it is not suitable to compare these formulas in silt range without revising. The expressions proposed in this paper can simulate the sediment concentration profiles for both silt and sand reasonably.

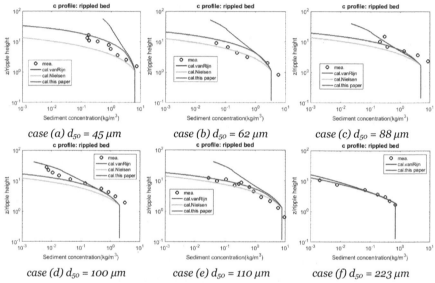

Figure 5-15. *Comparison of sediment concentration profile by different formulas (case (a): based on Li (2014)'s experiment with H = 0.18 m; case (b): based on Zhou and Ju (2007)'s experiment with H = 0.12 m; case (c): based on Yao et al. (2015)'s experiment with H = 0.10 m; case (d): based on Havinga (1992)'s experiment with H = 0.10 m; case (e): based on Li (2014)'s experiment with H = 0.18 m; case (f): based on van Rijn et al. (1993)'s experiment with H = 0.15 m and u_c = 0.13 m/s)*

5.4. Depth-averaged sediment concentration under waves

5.4.1. An expression for depth-averaged SSC in wave conditions

By integrating the SSC profile through water depth, an equilibrium depth-averaged concentration was yielded.

In 'flat bed' conditions, by integrating Eq. (5-11),

$$\overline{c_h} = \frac{1}{h}[\int_{z_a}^{z_1}\overline{c_a}(\frac{z}{z_a})^{-\alpha_f}dz + \int_{z_1}^{z_2}\overline{c_{z1}}(\frac{z-h'}{z}\frac{z_1}{z_1-h'})^{\frac{1}{a_f}\alpha_f}dz + \int_{z_2}^{h}\overline{c_{z2}}\exp(-\frac{\alpha_f}{\beta_s\delta_w}(z-z_2))dz] \quad (5\text{-}22)$$

Then, the depth-averaged SSC over 'flat bed' is

$$\overline{c_h} = \frac{1}{h}(\overline{c_af_1} + \overline{c_{z1}f_2} + \overline{c_{z2}f_3}) \quad (5\text{-}23)$$

in which,

$$f_1 = \begin{cases} \dfrac{1}{1-\alpha_{f1}}(z_a^{\alpha_{f1}}z_1^{1-\alpha_{f1}} - z_a) & \alpha_{f1} \neq 1 \\[2em] z_a \ln\dfrac{z_1}{z_a} & \alpha_{f1} = 1 \end{cases},$$

$$f_2 = \int_{z_1}^{z_2}(\frac{z_1}{z_1-h'}\frac{z-h'}{z})^{\frac{1}{a}\alpha_{f2}}dz, \text{ and } f_3 = (-\frac{\beta_s\delta_w}{\alpha_3})(\exp(-\frac{\alpha_{f3}}{\beta_s\delta_w}(h-z_2))-1).$$

$$\alpha_{fi} = \frac{\overline{w_{si}}}{\overline{\phi_{di}}\kappa u_*'} = \text{average Rouse number considering stratification effects}$$

over plan bed, i = 1, 2, 3.

$\overline{\phi_{di}}$ and $\overline{w_{si}}$ were calculated by the formulas of stratification effects and hindered settling respectively, from the averaged sediment concentration at each layer.

Over rippled bed, by integrating Eq. (5-19), yields,

$$\overline{c_h} = \frac{1}{h}(\overline{c_af_{0r}} + \overline{c_{z1r}f_{1r}} + \overline{c_{z2r}f_{2r}} + \overline{c_{z3r}f_{3r}}) \quad (5\text{-}24)$$

in which,

$$f_{0r} = (-\frac{\overline{\phi_d}\overline{\beta_0 v_{tN}}}{w_{s0}})(\exp(-\frac{w_{s0}}{\overline{\phi_d}\overline{\beta_0 v_{tN}}}(z_{1r}-z_a))-1),$$

$$f_{1r} = \begin{cases} \dfrac{1}{1-\alpha_{r1}}(z_{1r}^{\alpha_{r1}}z_{2r}^{1-\alpha_{r1}} - z_{1r}) & \alpha_{r1} \neq 1 \\[2em] z_{1r} \ln\dfrac{z_{2r}}{z_{1r}} & \alpha_{r1} = 1 \end{cases},$$

$$f_{2r} = \int_{z_{2r}}^{z_{3r}}(\frac{z_{2r}}{z_{2r}-h'}\frac{z-h'}{z})^{\frac{1}{a}\alpha_{r2}}dz, \text{ and } f_{3r} = \frac{d_h}{3(\alpha_d+1)}[1-(\frac{h-2\eta}{d_h})^{\alpha_{r3}+1}].$$

$$\alpha_{ri} = \frac{\overline{w_{si}}}{\overline{\phi_{di}}\beta_l\kappa u_{v*}} = \text{average Rouse number considering stratification effects}$$

over rippled bed, i = 0, 1, 2.

$$d_h = 4(h - 2\eta) - 3(z_{3r} - 2\eta) \quad \text{and} \quad \alpha_{r3} = \frac{\overline{w_s}}{\phi_d \kappa u_{v*}} \frac{h - 2\eta}{3b_{up}\eta}$$. The average value $\overline{\beta_i}$

at each layer was calculated by Eq. (5-18).

O'Donoghue et al. (2006)'s method was used for the criterion of bed forms, i.e., $\psi \geq 300$ is the sheet-flow regime. The rippled bed with lower steepness, i.e., $\eta / \lambda < 0.12$, is seen as dynamically plane and the sediment diffusion is calculated by 'flat bed' method with the enhanced roughness by the rippled bed form.

Then, combined Eq. (5-23) and Eq. (5-24), the synthetical depth-averaged SSC from ripple beds and sheet flow conditions in wave conditions is:

$$\overline{c_h} = \begin{cases} \dfrac{1}{h}(\overline{c_a}f_{0r} + \overline{c_{z1r}}f_{1r} + \overline{c_{z2r}}f_{2r} + \overline{c_{z3r}}f_{3r}) & \psi < 300 \quad and \quad \eta / \lambda \geq 0.12 \\[2mm] \dfrac{1}{h}(\overline{c_a}f_1 + \overline{c_{z1}}f_2 + \overline{c_{z2}}f_3) & \psi \geq 300 \quad or \quad \eta / \lambda < 0.12 \end{cases} \tag{5-25}$$

Appendix E shows the procedure of Eq. (5-23) and Eq. (5-24) in Matlab script.

Figure 5-16 shows the verification using some experimental datasets. It can be seen that the proposed depth-averaged SSC fits the measured data well, with the agreement being within a factor of ± 2 in most cases.

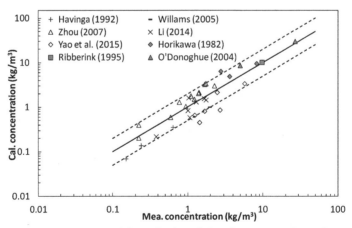

Figure 5-16. *Comparison of the calculated depth-averaged SSC by Eq. (5-25) with the measured data (The solid line indicates perfect agreement, and dashed lines indicate factor of ± 2 agreement)*

5.4.2. Discussion on the changes of sediment concentration under increasing wave dynamics conditions

Commonly, the bed form performs rippled bed in relatively low flow dynamics conditions, over which the sediment suspension is dominated by vortices; however, when the shear stress is further stronger, ripples are washed out, the effects of the vortex disappear and the roughness height becomes smaller. The vortices over rippled bed diffuse the sediment concentration to a higher position, which are more effective in transporting sediment than that above plane beds. Since rippled beds occur in relatively low flow dynamics conditions, this can lead to a paradoxical but logical outcome that, more sediment may be transported in the presence of small waves above rippled beds than by initial sheet flow beneath larger waves above plane beds (Davies and Villaret, 2002). Experiments by Yao et al. (2015) have shown that, for cases of s1-f3212 and s1-03812, ripples disappeared when currents were imposed on wave-only case s1-13, and the sediment concentration near the bottom decreased compared with the wave only case, even though the total shear stress is stronger. Some 1DV numerical models (e.g., Bijker, 1971; Davies and Villaret, 2002) exhibited the decrease in sediment transport rate and roughness when the initially rippled bed is washed out as the current strength increases. Therefore, the depth-averaged SSC is not always larger when flow dynamics is stronger.

To investigate this phenomenon, Figure 5-17 shows the depth-averaged sediment concentration (d_{50} = 62 μm) calculated by Eq. (5-25) with increasing wave dynamics, as well as the process of mobility number, wave shear stress, wave orbital velocity, ripple height and roughness height. It can be seen that the sediment concentration increases when wave shear stress is larger at the beginning, then decreases as the ripple height decreases; when the mobility number reaches the critical conditions of rippled bed to sheet flow, the sediment concentration is the smallest, because the effects of ripple disappear but the mobile bed is not fully-developed; from then on, the sediment concentration becomes higher when wave shear stress is stronger in sheet flow conditions. However, compared to the vortex rippled bed and sheet flow, the behavior of the turbulence and SSC in their transition zone (i.e., zone B) is still less studied, and more experimental data and theoretical works are needed.

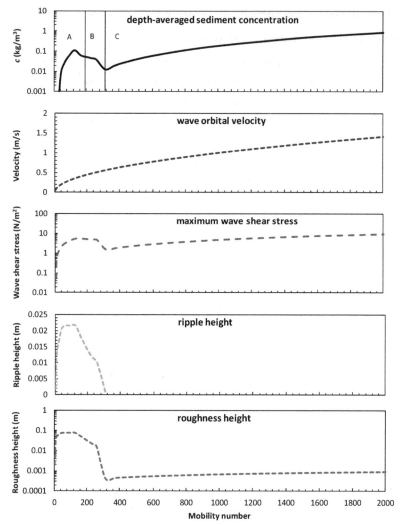

Figure 5-17. *The process of depth-averaged sediment concentration, wave orbital velocity, wave shear stress, and roughness height with mobility number (assumed conditions: h = 5 m, T = 5 s, d_{50} = 62 μm. zone A is over vortex rippled bed, zone B is over rippled bed but dynamically smooth and zone C is in sheet flow conditions)*

5.5. Conclusion

By solving the time-averaged diffusion equation for SSC and considering the effects of bed forms, stratification and hindered settling, expressions for time-averaged SSC profile under wave conditions was proposed for silt and are applicable for sand as well. Under combined wave-current conditions, numerical procedures were used for SSC profile. A number of datasets were collected for verification and reasonable results were obtained. The results are as follows:

Over 'flat bed', a three-layer distribution of wave-related eddy viscosity was proposed. The proposed SSC profile under wave conditions is power law in the low part and exponential distribution in the upper part, Eq. (5-11). Over rippled bed, a two-layer model was adopted (i.e., vortex-dominated layer and upper turbulence suspension layer). The proposed SSC profile under wave conditions is an exponential distribution in the vortex-dominated layer near bottom and power distributions at upper part, Eq. (5-19). For rippled bed with low steepness (<0.12), which is hydrodynamically plane, is treated as 'flat bed', but the roughness height is still calculated by the rippled bed method.

Afterwards, the depth-averaged sediment concentration was yielded by integrating the SSC profile under wave conditions, Eq. (5-25). The proposed formulas can describe the phenomenon that SSC does not always increase when wave dynamics increase due to effects of bed forms.

Sediment suspension is a complex physical process, which is impacted by many factors. For example, in natural environments the mixtures of clay, silt and sand would affect sediment suspension. The effects of bed forms on SSC are complicated, especially in the transition zone from rippled bed to sheet flow, where the sediment suspension is far more deeply understood; more measured data and research are needed for the turbulence process, sediment diffusivity and roughness etc.

Chapter 6

Conclusions and recommendations

This chapter presents the answers to the research questions and an overall summary of the main findings of this thesis. Recommendations on future study are discussed with respect to the breaking waves, sediment mixture and application to the 2D/3D models; furthermore the limitations of the results are mentioned.

6.1. Conclusions

This study aims to assist in better understanding silty sediments movement under wave-current actions. To analyze general sediment phenomena in silt-dominated coasts, field observations (chapter 2) was carried out in northwestern Caofeidian sea area of Bohai Bay, China, as well as other field data collection; laboratory experimental data analysis and theoretical analysis were used to study the incipient motion of silt-sand (chapter 3); a process based intra-wave 1DV model (chapter 4) was developed to simulate the flow-sediment dynamics near the bed bottom in combined wave-current conditions, in particular the HCL in wave-dominated conditions; based on verification using a number of experimental datasets, sensitivity analysis was carried out by the 1DV model on factors that impact the sediment concentration in HCL (chapter 4); finally, inspired by the 1DV model, the formulations of the mean sediment concentration profile of silty sediments were studied in purpose of practical application and 2D/3D simulation (chapter 5).

6.1.1. Answering the research questions

(1) How to develop the threshold criterion for silt, considering the differences and similarities between waves and currents, coarse and fine sediment?

The concept of the threshold for particle motion is widely accepted for both cohesive and non-cohesive sediments, i.e., sediment movement occurs when instantaneous fluid forces (entraining forces) on a particle (or an aggregate) are just larger than the instantaneous resisting forces (stabilizing forces). It is generally accepted that the entraining forces on a sediment grain could be adequately represented by the maximum shear stress generated by a flow, whether this flow is steady (current) or unsteady (wave with or without current).

Theoretical and force analysis were done from a unify perspective on sediment threshold motion. The analysis includes two aspects: the similarities and differences among the effects of waves and currents, and fine and coarse sediment. Based on previous knowledge, the stabilizing force includes the immersed gravity, the cohesive force, and the additional static water pressure. The submerged gravity is the main force when $d > 0.5$ mm; the cohesive force and additional static water pressure are the main forces when $d < 0.03$ mm; the gravity and cohesion are both important when $0.03 < d < 0.5$ mm. Compared with currents, the wave inertia-force is added in wave condition;

however, the wave inertia-force can be ignored, which makes the driving forces similar in uniform flow, waves, and combined wave-current conditions. As the consolidation process is slower when the grain size is smaller, the bulk density would be more important for silt than it is for coarse sand. Considering all stabilizing forces and driving forces, a unified view of the resisting forces for silt and sand sediments could be achieved.

Then, the formulation for sediment incipient motion of silt-sand was derived, following the route of the Shields curve. The Shields number was revised by adding the cohesive force and additional static pressure. The compaction coefficient is applied in a preliminary way and discussed to reflect the effect of bulk density for fine sediments.

(2) What are the key approaches for fine sediment modelling in BBL?

Based on the physical background, special approaches for silty sediment movement were considered during the 1DV model development, including bed forms (rippled bed and 'flat-bed'), hindered settling, stratification effects, mobile bed effects, reference concentration and critical shear stress.

Bed forms have great effects on turbulence and sediment entrainment. For silt, the effects of bed forms are much more important since the bed forms transform easily. Above plane beds, momentum transfer and sediment suspension occurs primarily by turbulent diffusion, while over rippled beds, momentum transfer and the associated sediment dynamics in the near-bed layer are dominated by coherent motions. The term 'plane bed' or 'flat bed' is used to refer to 'dynamically plane' bed, including sheet flow and rippled bed with mild steepness (<0.12). The normal k-ε turbulence model is employed for plane bed. The mobile bed effects caused by sheet flow layer have to be included for fine sediment transport modelling, which enhance the bed roughness. For finer sediments, the grain roughness becomes smaller and the enhanced roughness from mobile bed effects is dominated. For rippled beds, the combined vortex and k-ε model was employed to simulate the flow dynamics and the values of k and ε at the interface of the vortex-dominated layer were derived; the time series of the vortex and pick up function were given by the expressions of Davies and Thorne (2005), which describe the maximum eddy viscosity and reference sediment concentration at nearly the times of flow reversal. Different approaches have to be employed to simulate the sediment concentration over different bed forms. The method which simply generalizes the ripples based on roughness cannot correctly simulate the sediment dynamics.

The sediment-induced turbulence damping is an important term for high concentration layer modelling, i.e., the stratification effects. If the sediment concentration gradient is high, sediment-induced turbulence damping can largely affect the velocity profile and the transport rate, especially for fine sediment. The stratification effect was considered by adding the buoyancy flux in the k-ε model and employing the van Rijn (2007a)'s formula for the damping effect of vortex viscosity.

The settling velocity is significantly reduced when the sediment concentration is high, which is the so-called hindered settling effect. The approaches of hindered settling were employed considering the difference between silt and sand. Richardson and Zaki (1954)'s formula is employed for sand, and Te Slaa et al. (2015)'s formula is employed for silt.

Based on the results of chapter 3, the general expression of silt-sand incipience of motion was employed for the critical shear stress, which enables the model to be able to judge silt and sand threshold criterion in one procedure.

For the reference concentration on plane bed, it is unrealistic to give a zero value at the stage when the shear stress is lower than the critical value. To overcome this shortcoming, the reference concentration was revised by considering the deposited sediment from the last time step.

(3) What's the relationship between HCL and BBL? What are the main impact factors of SSC profile?

The formation of the HCL is strongly related to the turbulence production in BBL. In wave-dominated conditions, the turbulence is restricted in the BBL, which is one of the reasons of the HCL formation. The BBL is a combined result of bed forms and flow dynamics; we could directly establish the relation between the HCL and the BBL. The thickness of the HCL is about twice the height of the wave boundary layer, which means that, although the velocity is restricted in the BBL, the eddy viscosity as well as the diffusion viscosity can still reach higher levels. It is not appropriate to relate the HCL with a single factor, e.g., simple ripple parameters or flow parameters.

Sensitivity analysis was carried out by the 1DV model on the factors impacting the HCL, such as bed forms, flow dynamics, and effects of stratification and mobile bed. Bed forms determine the shape of the SSC profile near the bottom, and flow dynamics determine the magnitude. This is mainly caused by the difference of sediment diffusivity over different bed forms. Under sheet flow conditions, the maximum concentration near the bottom happens nearly at the phase of maximum flow shear dynamics. However, over rippled bed, it happens at the time of flow reversal because of the effects of the vortex. Bed forms are as important as flow dynamics for sediment transport, which means that we should not only analyze sediment transport with flow dynamics but also consider bed forms.

The stratification effects impact the sediment concentration greatly, which decrease the turbulence and concentration profile. The finer sediment grain size has more significant stratification effects. The stratification decreases the eddy viscosity (or sediment diffusivity), and shows a collapsing behaviour. The damping of turbulence contributes to the formation of the HCL, as the decreased diffusivity cannot sustain the sediment suspension. However, near the bottom, the damping effects show little change, which shows the maintenance of the turbulence production. This is mainly because there is no flocculation process and the bottom consists of a consolidated layer. The stratification effects of silt has the transitional behaviour between sand and cohesive mud, i.e., unlike sand, the stratification effects cannot be neglected; unlike fluid mud, the stratification effects for silt is not strong enough to destroy the flow dynamics.

For finer sediment, the mobile bed effects are larger because of a higher transport rate, i.e., the grain roughness is very low and the mobile bed roughness is dominant. Taking the mobile bed roughness into consideration, the sediment concentration will increase.

(4) How to develop time-averaged approaches for fine sediment modelling from BBL and HCL study for practical purposes (e.g., to improve the approaches of sediment simulation in 2DH/3D model)?

It is a challenge to apply the fundamental study results into practice. Normally, in morphological modelling, only the time-averaged value of SSC is needed. Based on the theoretical study and the 1DV model, we try to parameterize the process-based intra-wave results for practice purpose. Firstly, inspired by the 1DV model, distributions of wave-related sediment diffusivity over different bed forms were proposed; then, the time-averaged diffusion equation for suspended sediment was analytically solved by considering several important physical processes discussed in chapter 4, such as the effects of bed forms, stratification and hindered settling. The expressions for time-averaged SSC profile under wave conditions were proposed for silt and are applicable for sand as well. A number of datasets were collected for verification and reasonable results were obtained.

Over 'flat bed', a three-layer distribution of wave-related eddy viscosity was proposed. The proposed SSC profile under wave conditions is a power law in the low part and an exponential distribution in the upper part. Over rippled bed, a two-layer model was adopted (i.e., vortex-dominated layer and upper turbulence suspension layer). The proposed SSC profile under wave conditions is an exponential distribution in the vortex-dominated layer near bottom and power distributions at upper part. For rippled bed with low steepness (<0.12), which is hydrodynamically plane, is treated as 'flat bed', but the roughness height is calculated by the rippled bed method. The parameterization of SSC profiles is expected to be used for sediment concentration bottom boundary conditions in 3D modelling.

Afterwards, the depth-averaged sediment concentration was yielded by integrating the SSC profile under wave conditions, which is expected to be used for 2DH modelling as the source/sink term.

6.1.2. Overall conclusions

The main findings in this thesis are summarized as follows:

(1) Field observation data in several silt-dominated coasts shows that silt-dominated sediments are sensitive to flow dynamics. Under light flow dynamics the SSCs are normally small; however, the SSCs increase rapidly under strong flow dynamics (i.e., waves or strong tidal currents which can stirred up sediments). Meanwhile, because it is easy to settle down, the high concentration caused heavy sudden back siltation in navigation channels.

(2) A general expression of sediment incipient motion is proposed for silty to sandy sediments under the combined wave and current conditions. The Shields number was revised by adding the cohesive force and additional static pressure, which indicates that this study is an extension of Shields curve. The compaction coefficient is applied in a preliminary way and discussed to reflect the effect of bulk density for fine sediments. Verification and discussion are presented about the dependence of the critical shear stress on sediment particle size and bulk density, which shows satisfactory results.

(3) A process based intra-wave 1DV model was developed for flow dynamics and sediment transport in the wave-current bottom boundary layer, especially for simulation of the HCL of silty sediments. Based on the physical background, special approaches for sediment movement were considered in the model, including approaches for different bed forms, hindered settling,

stratification effects, mobile bed effects, reference concentration and critical shear stress. A number of experimental datasets were used to verify the model, which showed that the model is able to simulate the flow dynamics and sediment profiles well, including sheet flow conditions and rippled bed, as well as silt and sand. The simulation of the HCL helps us better understand the vertical concentration distributions of silt-dominated sediment under different wave-current conditions. It is a supplemental tool to flume experiments and a forerunner of 3D simulations. Meanwhile, it can serve as a simple reference model to test theoretical formulas, or to help assess the empirical parameterizations in those formulas for 2DH/3D modelling. Figure 6-1 and Figure 6-2 show the results of the 1DV model, i.e., the intra-wave process of velocity profile, eddy viscosity profile and sediment concentration profile over rippled bed and plane bed. The calculation conditions were based on Table 4-7.

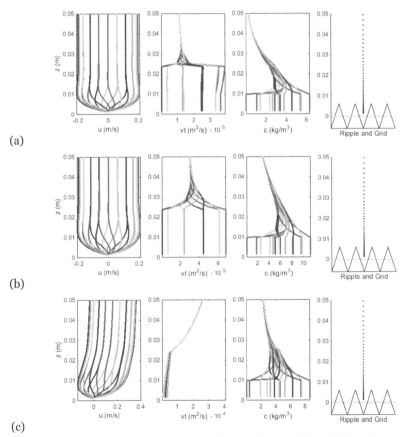

(a)

(b)

(c)

Figure 6-1. Intra-wave process of velocity profiles (left column), eddy *viscosity profiles (middle column) and sediment concentration profiles (right column) over rippled bed calculated by the 1DV model (The calculation conditions were based on d1-2 in Table 4-7, h=0.3 m, u_m = 0.2 m/s, T = 3 s and d_{50} = 0.062 mm. (a): wave only case with stratification effects; (b): wave only case without stratification effects; (c): wave-current case with u_c = 0.2 m/s)*

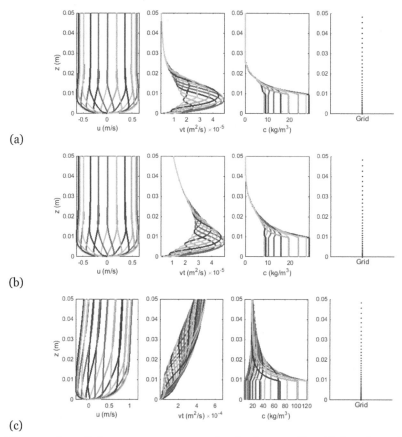

(a)

(b)

(c)

Figure 6-2. *Intra-wave process of velocity profiles (left column), eddy viscosity profiles (middle column) and sediment concentration profiles (right column) in sheet flow conditions calculated by the 1DV model (The calculation conditions were based on d1-9 in Table 4-7, h=0.3 m, u_m = 0.6 m/s, T = 3 s and d_{50} = 0.062 mm. (a): wave only case with stratification effects; (b): wave only case without stratification effects; (c): wave-current case with u_c = 0.6 m/s)*

(4) Sensitivity analysis was carried out on the factors impacting the HCL, i.e., bed forms, flow dynamics, and effects of stratification and mobile bed. The results showed that 1) the HCL is affected by both flow dynamics and bed forms; 2) as the BBL is the combined result of bed forms and flow dynamics, we could directly establish the relation between the HCL and the BBL; and 3) the thickness of the HCL is about twice the height of the wave boundary layer. Bed forms determine the shape of the concentration profile near the bottom, and flow dynamics determine the magnitude. Different approaches have to be employed to simulate the sediment concentration over different bed forms. The stratification effects impact the sediment concentration greatly and more works need to be carried out to investigate the stratification behaviour of silt (see Figure 6-1 and Figure 6-2). For finer sediment, the mobile bed effects are larger because of a higher transport rate, i.e., the grain roughness is very low and the mobile bed roughness is dominant. When a current is imposed, diffusion becomes larger in the upper part, the SSC profile becomes smoother, which means that the HCL normally exists in wave-dominant conditions.

(5) By solving the time-averaged diffusion equation for SSC and considering the effects of bed forms, stratification and hindered settling, expressions for time-averaged SSC profile under wave conditions was proposed for silt and are applicable for sand as well. Afterwards, the depth-averaged sediment concentration was yielded by integrating the SSC profile under wave conditions.

6.2. Recommendations

(1) The finer part of the silt may behave like clay, e.g., eroded in aggregates and the incipience of motion is greatly affected by floc size and the consolidation process. Because flocculation was not considered in this study, the formulations for threshold were not suitable for fine sediments with grain sizes less than 10-20 μm, for which the clay theory has to be employed.

(2) In natural environment, there are generally mixtures of clay, silt and sand. Bed composition will have effects on bed forms and sediment concentration. At this stage, this study only focuses on uniform sediment. The influence of sediment gradation is another important factor in some cases, i.e., the hiding-exposure effects between different sediment sizes, which has been studied extensively and is a future study direction for simulation of mixed sediment.

(3) The approaches in the 1DV model for rippled bed employed in this paper are based on empirical expressions, which are mainly suitable for

wave-dominated conditions. More sophisticated approaches are needed for further study.

(4) There are still many other parameters effecting sediment movement, e.g., mineralogy, organic content, and amounts and sizes of gas bubbles, which may need further study.

(5) The effects of bed forms on SSC are complicated, especially in the transition zone from rippled bed to sheet flow, where the sediment suspension is far more deeply understood; more measured data and research are needed for the turbulence process, sediment diffusivity, roughness etc.

(6) The parameterization study of SSC profile is in non-breaking waves. In surf zone, breaking waves exert great influence on silty sediment transport, which is a further study direction. Furthermore, it still needs to study the effects of non-linear wave characteristics on silty sediment transport, such as wave asymmetry, progressive wave streaming, acceleration skewness, phase-lag effects, etc.

(7) Under sheet flow conditions, our model focuses on the above suspension layer and the diffusion equation is used. In sheet flow layer, the sediment movement is supported by intergranular forces rather than the random turbulence; the flow will be non-Newton flow. If we penetrate into the sheet flow layer, the framework has to be changed, such as the approach of a pick-up function and a two-phase model.

(8) The formulations proposed in this study (i.e., the formulas of sediment incipient motion, the mean SSC profile and the depth-averaged SSC) are expected to be applied into practice in future, such as the morphological models. The expression of incipient motion can be used for critical shear stress in 2D/3D models, which has already been used in the 1DV model. The mean SSC profile can be used for bottom boundary conditions in 3D case, e.g., to transfer the source/sink terms at the reference height to the lowest computational cell, which depends on the assumption of the concentration distribution near the bed. The depth-averaged sediment concentration is expected to be used as the source term in 2DH case. It needs more calibration, verification, and extension to the waves-plus-current case.

Selected notation

a	Coefficient	u_{*w}	Wave maximum shear velocity
A	Wave amplitude	u_∞	Horizontal free stream velocity
BBL	Bottom boundary layer	υ	Velocity in y direction
B_k	Buoyancy flux	w	Vertical velocity
c	Sediment concentration by mass or volume	w_s	Settling velocity
c_a	Reference sediment concentration	$w_{s,0}$	Settling velocity in clear fluid
C_d	Drag force coefficient	W	Gravity force (submerged particle weight)
c_e	Wave celerity	z	Vertical distance above the bottom
c_{gel}	Gelling volume concentration	z_0	The height where velocity is zero
$c_{gel,s}$	Maximum volume concentration of sand bed	z_a	Reference height
$\overline{c_h}$	Mean depth-averaged sediment concentration	z_{max}	Maximum z in the calculation domain
C_L	Lift force coefficient	α	Rouse number
c_υ	Volume sediment concentration of solids	α_f	Rouse number over flat bed considering stratification effects
$c_{v,structure}$	Maximum volume fraction of solids	α_r	Rouse number over rippled bed considering stratification effects
c_*	Depth-averaged sediment transport capacity	β	Adjusted parameter of sediment diffusivity
c_μ	Coefficient in $k\text{-}\varepsilon$ model (= 0.09)	β_c	Compaction coefficient
d	Diameter of bed material	χ	Correction parameter
d_{50}	Median size of sediment	δ	Thickness of laminar layer near wall
d_{10}	Grain size for which 10% of the bed material is finer	δ_s	Thickness of bound water
d_{25}	Grain size for which 25% of the bed material is finer	δ_w	Thickness of wave boundary layer
d_s	Sieve diameter	ε	Turbulent dissipation

d_{silt}	$= 32\ \mu m$	ε_k	Cohesion coefficient
d_{sand}	$= 62\ \mu m$	ε_s	Sediment diffusivity
D	Deposition rate	$\varepsilon_{s,c}$	Current-related sediment diffusivity coefficient
D_*	Dimensionless particle size	$\varepsilon_{s,w}$	Wave-related sediment diffusivity coefficient
E	Erosion rate	$\varepsilon_{s,cw}$	Combined sediment diffusivity coefficient of waves and currents
F_c	Cohesive force	ε_{vortex}	Turbulent dissipation at the edge of vortex layer
F_D	Drag force	ϕ	Angle between wave and current
F_L	Lift force	ϕ_d	Damping coefficient
F_s	Source/sink term	$\phi_{cohesive}$	Cohesive effects coefficient
F_V	Wave inertia force	$\phi_{packing}$	Packing effects coefficient
f_w	Wave friction coefficient	$\phi_{s,max}$	Maximum density
F_δ	Additional static water pressure	$\phi_{s,struct}$	Structural density
g	Gravitational acceleration	η	Ripple height
h	Water depth	κ	Karman number
H	Wave height	λ	Ripple length
H_s	Significant wave height	θ	Shields number
HCL	High concentration layer	θ_c	Critical Shields number
J	Water surface slope	θ_{zc}	Revised critical Shields number (Incipience number)
k	Turbulent kinetic energy	θ_r	Ripple-adjusted value of Shields number
k_s	Roughness height	ρ	Water density
k_{vortex}	Turbulent kinetic energy at the edge of vortex layer	ρ_s	Sediment density
N	Brunt-Vaisala frequency	ρ_m	Density of water-sediment mixture
p	Water pressure	ρ_0	Dry density
$\overline{p}$	Time-averaged water pressure	$\rho_0{}^*$	Stable dry density

$\tilde{p}$	Oscillatory water pressure	ρ'	Wet density
Re_*	Shear Reynolds number	ρ'^*	Stable wet density
$\mathrm{Re}\,d_*$	Non-dimensional sediment Reynolds number	σ	Prandtl-Schmidt number
Re_{wave}	Wave Reynolds number	τ	Shear stress
s	Relative density	τ_c	Critical shear stress
SSC	Suspended sediment concentration	τ_{cr0}	Critical shear stress from Shields curve
T	Wave period	τ_{Re}	Reynolds stress
u	Instantaneous horizontal velocity	$\tau_{wc\max}$	Maximum shear stress beneath combined wave and current
$\bar{u}$	Mean horizontal velocity	τ_{wcmean}	Mean shear stress beneath combined wave and current
$\tilde{u}$	Oscillatory horizontal velocity	τ_{wm}	Maximum wave shear stress
u_o	The flow velocity near the sediment particle	υ	Kinematic viscosity coefficient
$u_{o,cr}$	The critical velocity on the particles near the bed	υ_t	Eddy viscosity
u_c	Mean current velocity	$\upsilon_{t,c}$	Current-related eddy viscosity
u_m	Maximum wave orbital velocity	$\upsilon_{t,w}$	Wave-related eddy viscosity
u_w	Wave-related velocity	υ_{tN}	Eddy viscosity in vortes-dominated layer
u_{wc}	Velocity of combined wave-current	ψ	Mobility number
u_*	Shear velocity	ω	Wave frequency
u_*'	Effective mean shear velocity	ζ	Water level

General operators

$^-$	Time-average, steady component
$^\sim$	Periodic component
$'$	Random component

Appendix A. Derivation of Reynolds equation for wave-current motions

From the simplification of the N-S equations, the governing equations in x-z coordinate are:

Momentum equation:

$$\frac{\partial u}{\partial t} + u\frac{\partial u}{\partial x} + w\frac{\partial u}{\partial z} = -\frac{1}{\rho}\frac{\partial p}{\partial x} + \upsilon(\frac{\partial^2 u}{\partial x^2} + \frac{\partial^2 u}{\partial z^2}) \tag{A.1}$$

Continuity equation:

$$\frac{\partial u}{\partial x} + \frac{\partial w}{\partial z} = 0 \tag{A.2}$$

where, u, w are instantaneous velocities in the x and z directions, respectively, p is the pressure, ρ is the water density, and υ is the kinetic viscosity.

Following the Reynolds' decomposition method, the Reynolds equations for wave-current BBL are obtained by splitting the variables into a fluctuating component u', an averaged component $\bar{u}$ and an oscillatory component $\tilde{u}$.

$$u = \bar{u} + \tilde{u} + u', \quad w = \bar{w} + \tilde{w} + w', \quad p = \bar{p} + \tilde{p} + p' \tag{A.3}$$

The oscillatory velocity $\tilde{u}$ is defined as: $\tilde{u} = \frac{1}{N}\sum_{j=1}^{N} u(z, t+jT) - \bar{u}(z)$, where

T is the wave period.

Taking the time-average, we obtain the momentum equation for $\bar{u}$

$$\frac{\partial \bar{u}}{\partial t} + [\frac{\partial \overline{uu}}{\partial x} + \frac{\partial \overline{uw}}{\partial z}] + [\frac{\partial \overline{\tilde{u}\tilde{u}}}{\partial x} + \frac{\partial \overline{\tilde{u}\tilde{w}}}{\partial z}] + [\frac{\partial \overline{u'u'}}{\partial x} + \frac{\partial \overline{u'w'}}{\partial z}] =$$
$$-\frac{1}{\rho}\frac{\partial \bar{p}}{\partial x} + \upsilon[\frac{\partial}{\partial x}\frac{\partial \bar{u}}{\partial x}] + \upsilon[\frac{\partial}{\partial z}\frac{\partial \bar{u}}{\partial z}] \tag{A.4}$$

The mean Reynolds stress can be expressed as

$$-\overline{u'w'} = \bar{\upsilon}_t \frac{\partial \bar{u}}{\partial z} \tag{A.5}$$

in which $\bar{\upsilon}_t$ is the eddy viscosity.

Then we get the momentum equation for the mean velocity:

$$\frac{\partial \bar{u}}{\partial t} + \frac{\partial}{\partial x}(\overline{uu} + \overline{\tilde{u}\tilde{u}}) + \frac{\partial}{\partial z}(\overline{uw} + \overline{\tilde{u}\tilde{w}}) = -\frac{1}{\rho}\frac{\partial \bar{p}}{\partial x} + \frac{\partial}{\partial z}[(\upsilon + \bar{\upsilon}_t)\frac{\partial \bar{u}}{\partial z}] \tag{A.6}$$

in which $\frac{\partial}{\partial x}\overline{\tilde{u}\tilde{u}} + \frac{\partial}{\partial z}\overline{\tilde{u}\tilde{w}}$ is the wave-induced stress term, which is analogous to the familiar Reynolds stresses (Nielsen, 1992).

Taking the phase-average, we obtain the equation for $\tilde{u}$

$$\frac{\partial \tilde{u}}{\partial t} + \frac{\partial}{\partial x}(\overline{\tilde{u}\tilde{u}}) + \frac{\partial}{\partial x}(\widetilde{uu}) + \frac{\partial}{\partial z}(\widetilde{uw} + \overline{\tilde{u}\tilde{w}}) + \frac{\partial}{\partial x}(\widetilde{\tilde{u}\tilde{u}}) + \frac{\partial}{\partial z}(\widetilde{\tilde{u}\tilde{w}}) + \frac{\partial}{\partial x}(\widetilde{u'u'}) + \frac{\partial}{\partial z}(\widetilde{u'w'}) =$$
$$-\frac{1}{\rho}\frac{\partial \tilde{p}}{\partial x} + \upsilon(\frac{\partial^2 \tilde{u}}{\partial x^2} + \frac{\partial^2 (\tilde{u})}{\partial z^2}) \tag{A.7}$$

The oscillatory Reynolds stress is define as

$$-\widetilde{u'w'} = \tilde{\upsilon}_t \frac{\partial \tilde{u}}{\partial z},$$ (A.8)

Then we get

$$\frac{\partial \tilde{u}}{\partial t} + \frac{\partial}{\partial x}(\overline{\tilde{u}\tilde{u}} + \overline{\tilde{u}\tilde{u}}) + \frac{\partial}{\partial z}(\overline{\tilde{u}\tilde{w}} + \overline{\tilde{u}\tilde{w}}) + \frac{\partial}{\partial x}(\widetilde{\tilde{u}\tilde{u}}) + \frac{\partial}{\partial z}(\widetilde{\tilde{u}\tilde{w}}) =$$
$$-\frac{1}{\rho}\frac{\partial \tilde{p}}{\partial x} + [\frac{\partial}{\partial x}(\upsilon + \tilde{\upsilon}_t)\frac{\partial \tilde{u}}{\partial x} + \frac{\partial}{\partial z}(\upsilon + \tilde{\upsilon}_t)\frac{\partial \tilde{u}}{\partial z}]$$ (A.9)

Because $\widetilde{\tilde{u}\tilde{u}} = \widetilde{\tilde{u}\tilde{u}} - \overline{\tilde{u}\tilde{u}}$, $\widetilde{\tilde{u}\tilde{w}} = \widetilde{\tilde{u}\tilde{w}} - \overline{\tilde{u}\tilde{w}}$ (Nielsen, 1992), thus,

$$\frac{\partial}{\partial x}\widetilde{\tilde{u}\tilde{u}} = \frac{\partial}{\partial x}\widetilde{\tilde{u}\tilde{u}} - \frac{\partial}{\partial x}\overline{\tilde{u}\tilde{u}}$$ (A.10)

$$\frac{\partial}{\partial z}\widetilde{\tilde{u}\tilde{w}} = \frac{\partial}{\partial z}\widetilde{\tilde{u}\tilde{w}} - \frac{\partial}{\partial z}\overline{\tilde{u}\tilde{w}}$$ (A.11)

We finally get the momentum equation for oscillatory motion

$$\frac{\partial \tilde{u}}{\partial t} + \frac{\partial}{\partial x}(\overline{\tilde{u}\tilde{u}} + \overline{\tilde{u}\tilde{u}} + \tilde{u}\tilde{u} - \overline{\tilde{u}\tilde{u}}) + \frac{\partial}{\partial z}(\overline{\tilde{u}\tilde{w}} + \overline{\tilde{u}\tilde{w}} + \tilde{u}\tilde{w} - \overline{\tilde{u}\tilde{w}}) = -\frac{1}{\rho}\frac{\partial \tilde{p}}{\partial x} + \frac{\partial}{\partial z}[(\upsilon + \tilde{\upsilon}_t)\frac{\partial \tilde{u}}{\partial z}]$$ (A.12)

The time scale in the Reynolds equation is much smaller than the oscillatory period but much larger than turbulence. Then, ignore the fluctuating components, $U = \bar{u} + \tilde{u}$, $P = \bar{p} + \tilde{p}$, where U and P represent Reynolds averaged components which are the sums of the averaged values and the oscillatory values.

Let $\bar{\upsilon}_t \frac{\partial \bar{u}}{\partial x} + \tilde{\upsilon}_t \frac{\partial \tilde{u}}{\partial x} = \upsilon_t \frac{\partial(\bar{u} + \tilde{u})}{\partial x}$, in which $\upsilon_t = (\bar{\upsilon}_t \frac{\partial \bar{u}}{\partial x} + \tilde{\upsilon}_t \frac{\partial \tilde{u}}{\partial x}) / \frac{\partial(\bar{u} + \tilde{u})}{\partial x}$.

By combining the time-averaged (Eq. (A.6)) and phase-averaged equations (Eq. (A.12)), we get the Reynolds equations for the wave-current BBL, Eq. (A.13). To simplify, we rewrite U as u and P as p in the following equations.

Momentum equation

$$\frac{\partial u}{\partial t} + u\frac{\partial u}{\partial x} + w\frac{\partial u}{\partial z} = -\frac{1}{\rho}\frac{\partial p}{\partial x} + \frac{\partial}{\partial z}[(\upsilon + \upsilon_t)\frac{\partial u}{\partial z}]$$ (A.13)

Similarly, the continuity equation is

$$\frac{\partial u}{\partial x} + \frac{\partial w}{\partial z} = 0$$ (A.14)

Appendix B. Time varying functions of the eddy viscosity and the reference concentration (Davies and Thorne, 2005)

The time varying eddy viscosity is assumed to be given by the real part of the following expression,

$$\upsilon_{tN}(t) = \overline{\upsilon_{tN}} f(\omega t) \tag{B.1}$$

with $f(\omega t) = (1 + \varepsilon_0 + \varepsilon_1 e^{i\omega t} + \varepsilon_2 e^{2i\omega t})$ (B.2)

in which $\varepsilon_1 = |\varepsilon_1| e^{i\varphi_1}$, $\varepsilon_2 = |\varepsilon_2| e^{i\varphi_2}$ (B.3)

For asymmetrical wave motion, the asymmetry parameter $B = |u_2 / u_1|$ is defined at the edge of the wave boundary layer by

$$u_\infty = u_1 e^{i\omega t} + u_2 e^{2i\omega t} \tag{B.4}$$

$$|\varepsilon_1| = \begin{cases} 10B & B \le 0.1 \\ 1.0 & B \ge 0.1 \end{cases} \tag{B.5}$$

$$|\varepsilon_2| = \begin{cases} 1.0 & B \le 0.1 \\ 1.0 - \dfrac{40}{3}(B - 0.1) & 0.1 \le B \le 0.15 \\ 1/3 & B \ge 0.15 \end{cases} \tag{B.6}$$

$$\varepsilon_0 = |\varepsilon_1|^2 / (8|\varepsilon_2|) \tag{B.7}$$

Davies and Thorne (2005) found that the peak eddy viscosity occurs just before flow reversal. The following phase relationships are used for the components of the eddy viscosity in relation to the instant of flow reversal following the passage of the wave crest,

$$\varphi_2 = 2\varphi_1, \quad \varphi_1 = -\arccos(B) + \Delta\varphi \tag{B.8}$$

with the phase lead of the peak eddy viscosity before flow reversal corresponding to $\Delta\varphi = 4°$.

The time-varying reference concentration over rippled bed is

$$c_a(t) = \overline{c_a} \frac{-0.5((1+\varepsilon_0) + \varepsilon_1 e^{i\omega t} + \varepsilon_2 e^{2i\omega t})((1 + a_c e^{2i\omega t}) + c.c)}{((1+\varepsilon_0) + 0.25 A_c |\varepsilon_0|(e^{i(2\varphi_1 - 2\phi_c)} + c.c))} \tag{B.9}$$

where $\overline{c_a}$ is the mean reference sediment concentration. $A_c = 1$. The coefficient $a_c = A_c \exp(2i\varphi_c)$. c.c. denotes the complex conjugate. The phase angle φ_c is taken as $\varphi_c = \varphi_1 + 30 \times (\pi/180)$ and leads to the outcome that the predicted concentration maxima at the crest level occur somewhat before flow reversal.

Appendix C. The formula for reference concentration by van Rijn (2007a) and Yao et al. (2015)

Yao et al. (2015) extended the formula for reference concentration proposed by van Rijn (2007a) to silt range,

$$c_a = \beta_y (1 - p_{clay}) f_{silt} \frac{d_{50}}{z_a} \frac{T_*^{1.5}}{D_*^{0.3}} \tag{C1}$$

in which, $\beta_y = 0.015$ is an original empirical coefficient for sand, and Yao et al. (2015) extended it to silt by using $\beta_y = 0.118 D_*^{-0.7}$, with a maximum value of 0.118 and minimum value of 0.015. $f_{silt} = d_{sand} / d_{50}$ is the silt factor ($f_{silt} = 1$ for $d_{50} > d_{sand}$), and d_{sand} = 62 μm. p_{clay} is the percentage of clay material in the bed. $D_* = d_{50}[(s-1)g/\upsilon^2]^{1/3}$ is the dimensionless particle size. The reference height z_a is defined as the maximum value of half the wave-related and half the current-related bed roughness values $k_{s,c,r}$, with a minimum value of 0.01m.

$$k_{s,c,r} = \begin{cases} 150 f_{cs} d_{50} & \psi \leq 50 \\ (182.5 - 0.652\psi) f_{cs} d_{50} & 50 < \psi \leq 250 \\ 20 f_{cs} d_{50} & \psi > 250 \end{cases} \tag{C2}$$

$$k_{s,c,r} = 20 d_{silt} \qquad d_{50} < d_{silt}$$

in which $f_{cs} = (0.25 d_{gravel} / d_{50})^{1.5}$ and f_{cs} = 1 for $d_{50} <= 0.25 d_{gravel}$, and d_{gravel} = 0.002 m. The lower limit is $f_{cs} = 20 d_{silt}$ = 640 μm for particles <= 32 μm. The minimum of wave-related bed roughness height in the sheet flow regime will be 0.01m.

$T_* = (\tau'_{b,cw} - \tau_c)/\tau_c$, in which τ_c is the critical bed shear stress and $\tau'_{b,cw}$ is originally the time-averaged effective bed-shear stress under currents and waves. $\tau'_{b,cw} = \tau'_{b,c} + \tau'_{b,w}$, where $\tau'_{b,c} = \mu_c \alpha_{cw} \tau_{b,c}$ = effective current-related bed-shear stress, and $\tau'_{b,w} = \mu_w \tau_{b,w}$ = effective wave-related bed-shear stress, μ_w = wave-related efficiency factor; and α_{cw} = wave-current interaction factor (van Rijn 1993).

The current-related efficiency factor is defined as $\mu_c = f'_c / f_c$, with f'_c = grain-related friction coefficient based on d_{90}; and f_c = current-related friction coefficient based on predicted bed roughness values. $C' = 18\log(12h/3d_{90})$, $C = 18\log(12h/k_{s,c})$, $f'_c = 8g/C'^2$, and $f_c = 8g/C^2$. $\mu_w = 0.7/D_*$, with $\mu_{w,min}$ = 0.14 for $D_* \geq 5$ and $\mu_{w,max}$ = 0.35 for $D_* \leq 5$.

Apparent roughness $k_a / k_{s,c} = \exp(\gamma_k u_w / u_c)$ with maximum value of 10, where $\gamma_k = 0.8 + \phi - 0.3\phi^2$ and ϕ = angle between wave direction and current direction (in radians between 0 and π).

Wave-current interaction coefficient,

$$\alpha_{cw} = \left[\frac{\ln(90\delta_w / k_a)}{\ln(90\delta_w / k_{s,c})} \right]^2 \left[\frac{-1 + \ln(30h / k_{s,c})}{-1 + \ln(30h / k_a)} \right]^2 \quad , \quad \text{with} \quad \alpha_{cw} = \max(\alpha_{cw}, 1) \quad . \quad \delta_w =$$

thickness of wave boundary layer $= 0.36 A_\delta (A_\delta / k_{s,w,r})^{-0.25}$, and $A_\delta =$ peak orbital excursion.

Appendix D. Description of the 1DV model: Input parameters and sketch frame of the modules

Input parameters
#Basic parameters

nz	=	Grid length	
rddz	=	Stretching factor	default = 1.05
h	=	Whole water depth	(m)
hh	=	Calculation water depth	(m)
rks	=	Basic roughness length	(m)

#Wave parameters

H	=	Significant wave height	(m)
T	=	Wave period	(s)
adv	=	Consider advection term or not	1 = yes, 0 = no
wtype	=	Wave type (linear wave, Stokes wave or harmonic componets	1 = linear wave; 2 = Stokes wave; 3 = harmonic componets

#Current parameters

uc	=	Depth-averaged current velocity	(m/s)
hj	=	Slope	
wcangle	=	Wave-current angle	(°)
ucup	=	Current velocity at upper boundary	(m/s)

Numerical parameters

dt	=	Time step	(s)
relax	=	Relax parameter	default = 1
turtype	=	Turbulence model type	1 = algebraic model 21 = standard k-eps model 22 = NTM k-eps model
err	=	Iteration precision	Default = 1e-5

Bed form information

bedtype	=	The type of bed forms	0 = call bedform module 1 = flat bed, 2 = rippled bed
eta	=	Ripple height	0 = calculated by formula Non 0 = give true value (m)
hlmd	=	Ripple length	0 = calculated by formula Non 0 = give true value (m)

Sediment parameters

d50	=	Medium sediment grain size	(mm)
d10	=	Grain size finer than 10%	(mm)
rous	=	Sediment density	(kg/m³)
wso	=	Settling velocity	0 = calculated by formula non 0 = give value (m/s)
idws	=	Hindered settling or not	0 = no hindered settling 1 = with hindered settling
idca	=	Reference concentration formulas	1 = Zyserman's formula 2 = Nielsen's formula 3 = Yao's formula
ibuoy	=	Stratification effects or not	0 = no stratification effects 1 = with stratification effects
imobile	=	Mobile bed effects or not	0 = no mobile bed effects 1 = with mobile bed effects

The frame wrok of the modules

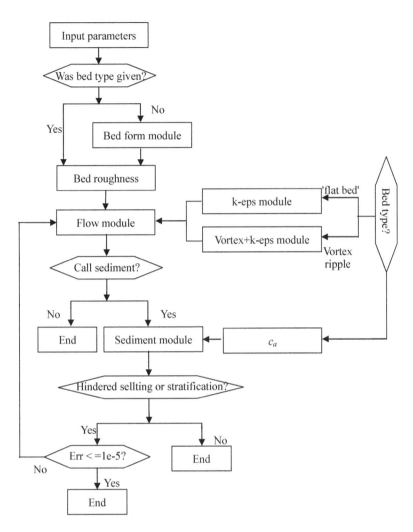

Figure D-1. *Sketch of the frame work of the 1DV model*

Appendix E. The procedures of depth-averaged SSC in Matlab script

The following scripts are for the depth-averaged SSC based on Eq. (5-23) and Eq. (5-24). All other scripts are available via the email of the author.

E1: The function of the depth-averaged SSC over 'flat bed', based on Eq. (5-23).

```
function[ch]=functionofcmeandepthaveragedflatbed(ca,za,h,dltaw
ave,uwstar,w0,d50,h1,h2)
    %ca:reference sediment concentration
    %za:reference height
    %h:water depth
    %dltawave:thickness of wave boundary layer
    %uwstar:wave shear velocity
    %w0:settling velocity in clear water
    %d50:median sediment grain size
    %h1:the low height of the cal.water colume,minmum za
    %h2:the high height of the cal.water colume,maxmum h
    karman=0.4;
    bup=0.8; %coefficent for diffusivity at upper part
    cc=2.5;  %upper layer
    dd=0.5;  %low layer
    a=bup/cc+(cc-bup)/(cc-dd); %coefficient
    b=(cc-bup)/(cc-dd);        %coefficient
    h1=max(h1,za);
    h2=min(h2,h);

    dampvanRijn1=1.0; %damping coefficient, initial
    dampvanRijn2=1.0; %damping coefficient, initial
    dampvanRijn3=1.0; %damping coefficient, initial
    w01=w0;           %initial
    w02=w0;           %initial
    w03=w0;           %initial
    z1=max(za,0.5*dltawave); %layer height
    z2=max(za,2.5*dltawave); %layer height
    hh=a*z2/b;               %coefficient
    ustarp=0.5*uwstar;       %half maximum shear velocity

    %do iteration, because of stratification effects and hindered
settling
    number=0;
    while (abs(ch-ch00)/ch>1.e-5)
        ch00=ch;
        %from za-z1
        zmin=max(za,h1);
        zmax=min(z1,h2);
        afa=1;
        if(zmax>zmin)
            afa=w01/karman/ustarp/dampvanRijn1; %Rouse number
            if(afa==1)
                f1=zmin*ln(zmax/zmin); % shape function f1
```

```
             else
                 f1=1/(1-afa)*(zmin^afa*zmax^(1-afa)-zmin);
             end
         end
         ch1=ca*f1/(zmax-zmin+1e-20); %c mean at layer za-z1
         %from z1-z2
         cz1=ca*(z1/za)^(-afa);           %c at level z1
         afa=w02/karman/ustarp/dampvanRijn2; %Rouse number
         zmin=max(z1,h1);
         zmax=min(z2,h2);
         if(zmax>zmin)
             nn2=20;
             ddz=(zmax-zmin)/nn2;
             f2=0;
             for i=1:nn2
                 z(i)=zmin+ddz*i;
             %shape function f2
                 f2=f2+ddz*(zmin/(zmin-hh)*(z(i)-hh)/z(i))^(afa/a);
             end
             ch2=cz1*f2/(zmax-zmin+1e-20); %c mean at layer z1-z2
         end
         %from z2-h
         cz2=cz1*((z2-hh)/z2*z1/(z1-hh))^(afa/a); %c at level z2
         afa=w03/karman/ustarp/dampvanRijn3;    %Rouse number
         zmin=max(z2,h1);
         zmax=min(h,h2);
         if(zmax>zmin)
         %shape function f3
             f3=-bup*dltawave/afa*(exp(-afa/bup/dltawave*
(zmax-zmin))-1);
             ch3=cz2*f3/(zmax-zmin+1e-20); %c mean at layer z1-z2
         end
         %final cmean
         ch=(ca*f1+cz1*f2+cz2*f3)/(h2-h1+1e-20); %depth-averaged c
mean btw h1 to h2

         %calculate the stratification effects
         dampvanRijn1=functionofdampvanRijn(ch1,d50);
         dampvanRijn2=functionofdampvanRijn(ch2,d50);
         dampvanRijn3=functionofdampvanRijn(ch3,d50);

         %hindered settling
         w01=functionofhinderedsettling(w0,d50,ch1);
         w02=functionofhinderedsettling(w0,d50,ch2);
         w03=functionofhinderedsettling(w0,d50,ch3);

         number=number+1;
     end
```

E2: The function of the depth-averaged SSC over rippled bed, based on Eq. (5-24).

```
function[ch]=functionofcmeandepthaveragedrippled(ca,za,h,eita,
vortex,w0,d50,h1,h2)
    %ca:reference sediment concentration
    %za:reference height
    %h:water depth
    %eita:ripple height
    %vortex:eddy vortex
    %w0:settling velocity in clear water
    %d50:median sediment grain size
    %h1:the low height of the cal.water colume,minmum za
    %h2:the high height of the cal.water colume,maxmum h
    karman=0.4;
    bup=2.25; %coefficient
    cc=4.5;   %upper layer
    dd=2.5;   %low layer
    a=bup/cc+(cc-bup)/(cc-dd);
    b=(cc-bup)/(cc-dd);
    h1=max(h1,za);
    h2=min(h2,h);

    dampvanRijn0r=1.0; %initial
    dampvanRijn1r=1.0; %initial
    dampvanRijn2r=1.0; %initial
    dampvanRijn3r=1.0; %initial
    w00=w0;            %initial
    w01=w0;            %initial
    w02=w0;            %initial
    w03=w0;            %initial
    %the height of each layer
    z1r=max(za,2*eita);    %vortex layer
    z2r=max(za,2.5*eita);  %
    z3r=max(za,4.5*eita);  %
    %average beita in each layer
    rr=1.0;
    beitav=4;
    rrr=3/(h-2*eita)^rr/(rr+1);
    beita1=rrr*((z2r-2*eita)^(rr+1)-(z1r-2*eita)^(rr+1));
    beita1=4-beita1/(z2r-z1r+1e-20);
    beita2=rrr*((z3r-2*eita)^(rr+1)-(z2r-2*eita)^(rr+1));
    beita2=4-beita2/(z3r-z2r+1e-20);
    beita3=rrr*((h-2*eita)^(rr+1)-(z3r-2*eita)^(rr+1));
    beita3=4-beita3/(h-z3r+1e-20);
    %efficiency shear velocity
    ustarp=0.5*vortex/karman/eita;
    %do iteration, because damp and w0 are affected by c
    hh=a*z3r/b; %coefficient
    number=0;
    while (abs(ch-ch00)/ch>1.e-5)
        ch00=ch;
        %from za to z1r
        zmin=max(za,h1);
        zmax=min(z1r,h2);
```

```
        if(zmax>zmin)
            f0r=(-dampvanRijn0r*beitav*vortex/w00)*(exp(-w00/
dampvanRijn0r/vortex/beitav*(zmax-zmin))-1);
            ch0=ca*f0r/(zmax-zmin+1e-20); %cmean from za-z1r
        end
        %from z1r-z2r
        %c at z1r
        cz1r=ca*exp(-w00/dampvanRijn0r/beitav/vortex*(z1r-za));
        afa=w01/karman/ustarp/dampvanRijn1r/beita1;
        zmin=max(z1r,h1);
        zmax=min(z2r,h2);
        if(zmax>zmin)
            if(afa==1)
                f1r=zmin*ln(zmax/zmin);
            else
                f1r=1/(1-afa)*(zmin^afa*zmax^(1-afa)-zmin);
            end
            ch1=cz1r*f1r/(zmax-zmin+1e-20);
        end
        %from z2r-z3r
        cz2r=cz1r*(z2r/z1r)^(-afa);
        afa=w02/karman/ustarp/dampvanRijn2r/beita2;
        zmin=max(z2r,h1);
        zmax=min(z3r,h2);
        if(zmax>zmin)
            nn2=20;
            ddz=(zmax-zmin)/nn2; %divided z2r-z3r by nn2 times
            f2r=0;
            for i=1:nn2
                z(i)=zmin+ddz*i;
                f2r=f2r+ddz*(zmin/(zmin-hh)*(z(i)-hh)/z(i))^
(afa/a);
            end
            ch2=cz2r*f2r/(zmax-zmin+1e-20);
        end
        %from z3r-h
        cz3r=cz2r*((z3r-hh)/z3r*z2r/(z2r-hh))^(afa/a);
        zmin=max(z3r,h1);
        zmax=min(h,h2);
        if(zmax>zmin)
            dh=4*(h-2*eita)-3*(z3r-2*eita);
            afad=w03/karman/ustarp/dampvanRijn3r;
            afad=afad*(h-2*eita)/3/bup/eita;
            f3r=dh/3/(afad+1)*(1-((h-2*eita)/dh)^(afad+1));
            ch3=cz3r*f3r/(zmax-zmin+1e-20);
        end
        %final cmean
        ch=(ca*f0r+cz1r*f1r+cz2r*f2r+cz3r*f3r)/(h2-h1+1e-20);

        %calculate the stratification effects
        dampvanRijn0r=functionofdampvanRijn(ch0,d50);
        dampvanRijn1r=functionofdampvanRijn(ch1,d50);
        dampvanRijn2r=functionofdampvanRijn(ch2,d50);
        dampvanRijn3r=functionofdampvanRijn(ch3,d50);
```

```
%hindered settling
w00=functionofhinderedsettling(w0,d50,ch0);
w01=functionofhinderedsettling(w0,d50,ch1);
w02=functionofhinderedsettling(w0,d50,ch2);
w03=functionofhinderedsettling(w0,d50,ch3);

number=number+1;
end
```

References

Ahmed, A.S. and Sato, S., 2003. A sheetflow transport model for asymmetric oscillatory flows: Part I: Uniform grain size sediments. Coastal Engineering Journal, 45(03): 321-337.

Amoudry, L.O. and Souza, A.J., 2011. Deterministic coastal morphological and sediment transport modeling: A review and discussion. Reviews of Geophysics, 49(2): 1-21.

Ariathurai, R. and Arulanandan, K., 1978. Erosion rates of cohesive soils. Journal of the hydraulics division, 104(2): 279-283.

Ariathurai, R. and Krone, R.B., 1976. Mathematical modeling of sediment transport in estuaries. Estuarine Processes, (2): 98-106.

Baas, J.H., Best, J.L. and Peakall, J., 2016. Predicting bedforms and primary current stratification in cohesive mixtures of mud and sand. Journal of the Geological Society, 173(1): 12-45.

Bagnold, R.A., 1963. Beach and nearshore processe-Part 1, Mechanics of marine sedimentation. Physics of Sediment Transport by Wind and Water, 188.

Bagnold, R.A. and Taylor, G., 1946. Motion of waves in shallow water. Interaction between waves and sand bottoms, Proceedings of the Royal Society of London. Series A. Mathematical and Physical Sciences, pp 1-15.

Bakker, W.T. and Van Doorn, T., 1978. Near-bottom velocities in waves with a current. Coastal Engineering Proceedings, 1(16): 1394-1413.

Baldock, T.E., Tomkins, M.R., Nielsen, P. and Hughes, M.G., 2004. Settling velocity of sediments at high concentrations. Coastal Engineering, 51(1): 91-100.

Beheshti, A.A. and Ataie-Ashtiani, B., 2008. Analysis of threshold and incipient conditions for sediment movement. Coastal Engineering, 55(5): 423-430.

Bijker, E.W., 1967. Some considerations about scales for coastal models with movable bed. Delft Hydraulics Laboratory Publication 50. Delft, the Netherlands.

Bijker, E.W., 1971. Longshore transport computations. Journal of the Waterways, Harbors and Coastal Engineering Division, 97(4): 687-701.

Bijker E. W., Hijum E. V., Vellinga P., 1976. Sand transport by waves. Proc 15th ICCE, Honolulu, pp 1149-1167.

Blumberg, A.F., 2002. A primer for ECOMSED. Mahwah NJ: Hydro Qual Inc: 1-194.

Bolaños, R., Thorne, P.D. and Wolf, J., 2012. Comparison of measurements and models of bed stress, bedforms and suspended sediments under combined currents and waves. Coastal Engineering, 62: 19-30.

Brown, C.B., 1950. Sediment transportation. John Wiley and Sons.

Cacchione, D.A., Grant, W.D., Drake, D.E. and Glenn, S.M., 1987. Storm - dominated bottom boundary layer dynamics on the Northern California Continental Shelf: Measurements and predictions. Journal of Geophysical Research: Oceans (1978–2012), 92(C2): 1817-1827.

Cacchione, D.A., Sternberg, R.W. and Ogston, A.S., 2006. Bottom instrumented tripods: History, applications, and impacts. Continental shelf research, 26(17): 2319-2334.

Camenen, B. and Larson, M., 2005. A general formula for non-cohesive bed load sediment transport. Estuarine, Coastal and Shelf Science, 63(1): 249-260.

Camenen, B., Larson, M. and Bayram, A., 2009. Equivalent roughness height for plane bed under oscillatory flow. Estuarine Coastal & Shelf Science, 81(3): 409-422.

Carstens, M.R., Neilson, F.M. and Altinbilek, H.D., 1969. Bed forms generated in the laboratory under an oscillatory flow: Analytical and experimental study. Georgia Inst of Tech Atlanta.

Cao, Z., Kong, L. and Jiao, G., 2003. Sediment incipient motion under co-action of waves and currents. Acta Oceanologica Sinica, 25(3): 113-119 (in Chinese with English abstract).

Cao, Z., Yang, S.S. and Yang, H., 2003. Definition of silt-sandy beach and its characteristics of sediment movement. Port & Waterway Engineering, 5: 1-5 (in Chinese with English abstract).

Christoffersen, J.B. and Jonsson, I.G., 1985. Bed friction and dissipation in a combined current and wave motion. Ocean Engineering, 12(5): 387-423.

Coleman, N.L., 1969. A new examination of sediment suspension in open channels. Journal of Hydraulic Research, 7(1): 69-82.

Conley, D.C., Falchetti, S., Lohmann, I.P. and Brocchini, M., 2008. The effects of flow stratification by non-cohesive sediment on transport in high-energy wave-driven flows. Journal of Fluid Mechanics, 610: 43-67.

Da Silva, P.A., Temperville, A. and Seabra Santos, F., 2006. Sand transport under combined current and wave conditions: A semi-unsteady, practical model. Coastal Engineering, 53(11): 897-913.

Davies, A.G. and Thorne, P.D., 2005. Modeling and measurement of sediment transport by waves in the vortex ripple regime. Journal of Geophysical Research: Oceans (1978–2012), 110(C5).

Davies, A.G. and Villaret, C., 1999. Eulerian drift induced by progressive waves above rippled and very rough beds. Journal of Geophysical Research: Oceans, 104(C1): 1465-1488.

Davies, A.G. and Villaret, C., 2002. Prediction of sand transport rates by waves and currents in the coastal zone. Continental Shelf Research, 22(18): 2725-2737.

Davies, A.G. and Villaret, C., 2003. Sediment transport modelling for coastal morphodynamics, Proceedings of coastal sediments. Citeseer, pp 18-23.

Davies, A.G., Soulsby, R.L. and King, H.L., 1988. A numerical model of the combined wave and current bottom boundary layer. Journal of Geophysical Research: Oceans (1978–2012), 93(C1): 491-508.

Davies, A.G., Van Rijn, L.C., Damgaard, J.S., Van de Graaff, J. and Ribberink, J.S., 2002. Intercomparison of research and practical sand transport models. Coastal Engineering, 46(1): 1-23.

De Vriend, H.J. et al., 1993. Medium-term 2DH coastal area modelling. Coastal Engineering, 21(1): 193-224.

Debnath, K., Nikora, V., Aberle, J., Westrich, B. and Muste, M., 2007. Erosion of cohesive sediments: Resuspension, bed load, and erosion patterns from field experiments. Journal of Hydraulic Engineering, 133(5): 508 - 520.

Deigaard, R., Jakobsen, J.B. and Fredsøe, J., 1999. Net sediment transport under wave groups and bound long waves. Journal of Geophysical Research Oceans, 104(C6): 13559–13575.

Deltares, 2014. Delft3D-FLOW, Simulation of multi-dimensional hydrodynamic flows and transport phenomena, including sediments. User Manual, Delft, Holanda.

Dibajnia, M. and Watanabe, A., 1992. Sheet flow under nonlinear waves and currents. Coastal Engineering Proceedings, pp 2015-2028.

Dissanayake, D., Wurpts, A., Miani, M., Knaack, H., Niemeyer, H.D., Roelvink, J.A., 2012. Modelling morphodynamic response of a tidal basin to an anthropogenic effect: Ley Bay, East Frisian Wadden Sea–applying tidal forcing only and different sediment fractions. Coastal Engineering, 67: 14-28.

Dohmen-Janssen, C.M., Hassan, W.N. and Ribberink, J.S., 2001. Mobile - bed effects in oscillatory sheet flow. Journal of Geophysical Research: Oceans (1978–2012), 106(C11): 27103-27115.

Dohmen-Janssen, C.M., Kroekenstoel, D.F., Hassan, W.N. and Ribberink, J.S., 2002. Phase lags in oscillatory sheet flow: Experiments and bed load modelling. Coastal Engineering, 46(1): 61-87.

Dong, L.P., Sato, S. and Liu, H., 2013. A sheetflow sediment transport model for skewed-asymmetric waves combined with strong opposite currents. Coastal Engineering, 71: 87-101.

Dou, G., 1962. Theory of incipient motion of sediment. Scientia Sinica, 11(7): 991-999 (in Russian).

Dou, G., 2000. Incipient motion of sediment under currents. China Ocean Engineering, 14(04): 391-406.

Dou, G., 2001. Similarity theory of total sediment transport modeling for estuarine and coastal regions. Hydro-Science and Engineering, (1): 1-12 (in Chinese with English abstract).

Dou, G., Dong, F. and Dou, X., 1995. Sediment transport capacity of tidal currents and waves. Chinese Science Bulletin, (13): 1096-1101.

Dou, G., Dou, X. and Li, T., 2001. Incipient motion of sediment by waves. Science in China Series E: Technological Sciences, 44(3): 309-318.

Drønen, N. and Deigaard, R., 2007. Quasi-three-dimensional modelling of the morphology of longshore bars. Coastal Engineering, 54(3): 197-215.

Eagleson, P.S., Dean, R.G. and Peralta, L.A., 1957. The mechanics of the motion of discrete spherical bottom sediment particles due to shoaling waves. Hydrodynamics Laboratory, Massachusetts Institute of Technology.

Einstein, H.A., 1950. The bed-load function for sediment transportation in open channel flows. US Department of Agriculture.

Engelund, F. and Hansen, E., 1972. A monograph on sediment transport in alluvial streams. Teknisk forlag Copenhagen.

Etemad-Shahidi, A., Kazeminezhad, M.H. and Mousavi, S.J., 2009. On the prediction of wave parameters using simplified methods. Journal of Coastal Research, Special Issue 56: 505-509.

Everts, C.H., 1973. Particle overpassing on flat granular boundaries. Journal of the Waterways, Harbors and Coastal Engineering Division, 99(4): 425-438.

Fain, A., Ogston, A.S. and Sternberg, R.W., 2007. Sediment transport event analysis on the western Adriatic continental shelf. Continental Shelf Research, 27(3): 431-451.

Fernandez Luque, R. and Van Beek, R., 1976. Erosion and transport of bed-load sediment. Journal of Hydraulic Research, 14(2): 127-144.

Ferrarin, C. et al., 2008. Development and validation of a finite element morphological model for shallow water basins. Coastal Engineering, 55(9): 716-731.

Fredsøe, J., 1984. Sediment transport in current and waves. Institute of Hydrodynamics and Hydraulic Engineering, Technical University of Denmark.

Fredsøe, J. and Deigaard, R., 1992. Mechanics of coastal sediment transport, 3. World scientific.

Galappatti, G. and Vreugdenhil, C.B., 1985. A depth-integrated model for suspended sediment transport. Journal of Hydraulic Research, 23(4): 359-377.

Gonzalez-Rodriguez, D. and Madsen, O.S., 2007. Seabed shear stress and bedload transport due to asymmetric and skewed waves. Coastal Engineering, 54(12): 914-929.

Grant, W.D. and Madsen, O.S., 1979. Combined wave and current interaction with a rough bottom. Journal of Geophysical Research: Oceans (1978–2012), 84(C4): 1797-1808.

Grant, W.D. and Madsen O.S., 1982. Movable bed roughness in unsteady oscillatory flow. Journal of Geophysical Research: Oceans (1978–2012), 87(C1): 469-481.

Grant, W.D. and Madsen, O.S., 1986. The continental-shelf bottom boundary layer. Annual Review of Fluid Mechanics, 18(1): 265-305.

Grant, W.D., Williams III, A.J. and Glenn, S.M., 1984. Bottom stress estimates and their prediction on the northern California continental shelf during CODE-1: The importance of wave-current interaction. Journal of Physical Oceanography, 14(3): 506-527.

Grasmeijer, B.T. and Kleinhans, M.G., 2004. Observed and predicted bed forms and their effect on suspended sand concentrations. Coastal Engineering, 51(5): 351-371.

Grass, A.J., 1970. Initial instability of fine bed sand. Journal of the Hydraulics Division, 96(3): 619-632.

Guizien, K., Dohmen Janssen, M. and Vittori, G., 2003. 1DV bottom boundary layer modeling under combined wave and current: Turbulent separation and phase lag effects. Journal of Geophysical Research: Oceans (1978–2012), 108(C1): 3016.

Hammond, T.M. and Collins, M.B., 1979. On the threshold of transport of sand - sized sediment under the combined influence of unidirectional and oscillatory flow. Sedimentology, 26(6): 795-812.

Hanson, H. and Camenen, B., 2007. Closed form solution for threshold velocity for initiation of sediment motion under waves, Proceedings of Coastal Sediments, pp 15-28.

Hassan, W. and Ribberink, J.S., 2010. Modelling of sand transport under wave-generated sheet flows with a RANS diffusion model. Coastal Engineering, 57(1): 19-29.

Havinga, F.J., 1992. Sediment concentrations and sediment transport in case of irregular non-breaking waves with a current. Delft University of Technology, Delft, the Netherlands.

Henderson, S.M., Allen, J.S. and Newberger, P.A., 2004. Nearshore sandbar migration predicted by an eddy - diffusive boundary layer model. Journal of Geophysical Research: Oceans (1978–2012), 109: C06024.

Hill, I.M.N., 1963. The Sea. Interscience Publishers, New York, pp 538-539.

Holmedal, L.E. and Myrhaug, D., 2006. Boundary layer flow and net sediment transport beneath asymmetrical waves. Continental Shelf Research, 26(2): 252-268.

Holmedal, L.E. and Myrhaug, D., 2009. Wave-induced steady streaming, mass transport and net sediment transport in rough turbulent ocean bottom boundary layers. Continental Shelf Research, 29(7): 911-926.

Holmedal, L.E., Myrhaug, D. and Eidsvik, K.J., 2004. Sediment suspension under sheet flow conditions beneath random waves plus current. Continental Shelf Research, 24(17): 2065-2091.

Hooshmand, A., Horner-Devine, A.R. and Lamb, M.P., 2015. Structure of turbulence and sediment stratification in wave-supported mud layers. Journal of Geophysical Research: Oceans, 120(4): 2430-2448.

Horikawa, K. and Watanabe, A., 1967. A study of sand movement due to wave action. Coastal Engineering in Japan, (10): 39-57.

Horikawa, K., Watanabe, A. and Katori, S., 1982. Sediment transport under sheet flow condition. Coastal Engineering Proceedings, 1(18): 1335-1352.

Hsu, T. and Ou, S., 1994. Mean sediment concentration and turbulent boundary layer of wave-induced sheet flow. Journal of Hydraulic Research, (5): 675-687.

Hsu, T.J., Ozdemir, C.E. and Traykovski, P.A., 2009. High - resolution numerical modeling of wave - supported gravity - driven mudflows. Journal of Geophysical Research: Oceans, 114: C05014.

Huang, J., 1989. An experimental study of the scouring and settling properties of cohesive sediment. The Ocean Engineering, 1: 61-70 (in Chinese with English abstract).

Israelachvili, J., 1992. Intermolecular and Surface Forces, 68. Academic press, San Diego, pp 577-578.

Jayaratne, M.P.R., Rahman, M.R. and Shibayama, T., 2015. A cross-shore beach profile evolution model. Coastal Engineering Journal, 5670(4): 56-101.

Jayaratne, M.P.R., Srikanthan, S. and Shibayama, T., 2011. Examination of the suspended sediment concentration formulae using full-scale rippled bed and sheet-flow data. Coastal Engineering Journal, 53(04): 451-489.

Jensen, B.J., Sumer, B.M. and Fredsoe, J., 1989. Turbulent oscillatory boundary layers at high Reynolds numbers. Journal of Fluid Mechanics, 206(206): 265-297.

Jiang, C.B., Bai, Y.C., Jiang, N.S. and Hu, S.X., 2001. Incipient motion of cohesive silt in the Haihe River estuary. Journal of Hydraulic Engineering, 32(6): 51-56.

Jonsson, I.G., 1966. Wave boundary layers and friction factors. Coastal Engineering (1966). ASCE, pp 127-148.

Kantardgi, I.G., 1992. Incipiency of sediment motion under combined waves and currents. Journal of Coastal Research, 8(2): 332-339.

Kapdasli, M.S. and Dyer, K.R., 1986. Threshold conditions for sand movement on a rippled bed. Geo-marine letters, 6(3): 161-164.

Katopodi, I., Ribberink, J.S., Ruol, P. and Lodahl, C., 1994. Sediment transport measurements in combined wave-current flows. Coastal Dynamics94, ASCE, pp 837-851.

Katori, S., Sakakiyama, T. and Watanabe, A., 1984. Measurement of sand transport in a cross unidirectional-oscillatory flow tank. Coastal Engineering in Japan, 27: 193-203.

Khelifa, A. and Ouellet, Y., 2000. Prediction of sand ripple geometry under waves and currents. Journal of Waterway, Port, Coastal, and Ocean Engineering, 126(1): 14-22.

Kineke, G.C., Sternberg, R.W., Trowbridge, J.H. and Geyer, W.R., 1996. Fluid-mud processes on the Amazon continental shelf. Continental Shelf Research, 16(5): 667-696.

King, C.A., 1972. Beaches and coasts. Edward Arnold Ltd, London.

King, D. B., 1991. Studies in oscillatory flow bedload sediment transport. PhD Thesis, Univ of California, San Diego.

Klopman, G., 1994. Vertical structure of the flow due to waves and currents: laser-Doppler flow measurements for waves following or opposing a current, Delft Hydraulic Laboratory Report H840 Part2.

Komar, P.D. and Miller, M.C., 1974. Sediment threshold under oscillatory waves. Coastal Engineering Proceedings, 1(14): 756-775.

Kothyari, U.C. and Jain, R.K., 2008. Influence of cohesion on the incipient motion condition of sediment mixtures. Water Resources Research, 44(4): W04410.

Kranenburg, W.M., Ribberink, J.S., Schretlen, J.J. and Uittenbogaard, R.E., 2013. Sand transport beneath waves: the role of progressive wave streaming and other free surface effects. Journal of Geophysical Research: Earth Surface, 118(1): 122-139.

Kranenburg, W.M., Ribberink, J.S., Uittenbogaard, R.E. and Hulscher, S.J., 2012. Net currents in the wave bottom boundary layer: On waveshape streaming and progressive wave streaming. Journal of Geophysical Research, 117: F03005.

Krone, R.B., 1962. Flume studies of the transport of sediment in estuarial shoaling processes. University of California, USA.

Kuang, C. et al., 2012. A two-dimensional morphological model based on a next generation circulation solver I: Formulation and validation. Coastal Engineering, 59(1): 1-13.

Lamb, M.P. and Parsons, J.D., 2005. High-density suspensions formed under waves. Journal of Sedimentary Research, 75(3): 386-397.

Lee-Young, J.S. and Sleath, J., 1989. Initial motion in combined wave and current flows. Coastal Engineering 1988, 1140-1151.

Lesser, G.R., Roelvink, J.A., Van Kester, J. and Stelling, G.S., 2004. Development and validation of a three-dimensional morphological model. Coastal Engineering, 51(8): 883-915.

Li, M., Fernando, P.T., Pan, S., O'Connor, B.A. and Chen, D., 2007. Development of a quasi-3d numerical model for sediment transport prediction in the coastal region. Journal of Hydro-environment Research, 1(2): 143-156.

Li, M.Z. and Amos, C.L., 1999. Sheet flow and large wave ripples under combined waves and currents: field observations, model predictions and effects on boundary layer dynamics. Continental Shelf Research, 19(5): 637-663.

Li, M.Z. and Amos, C.L., 2001. SEDTRANS96: the upgraded and better calibrated sediment-transport model for continental shelves. Computers & Geosciences, 27(6): 619-645.

Li, S., 2014. Turbulent structure and sediment alluvial process in wave-current boundary layer, Nanjing Hydraulic Research Institute, Nanjing (in Chinese with English abstract).

Li, Z. and Davies, A.G., 1996. Towards predicting sediment transport in combined wave-current flow. Journal of Waterway, Port, Coastal, and Ocean Engineering, 122(4): 157-164.

Liang, B., Li, H. and Lee, D., 2007. Numerical study of three-dimensional suspended sediment transport in waves and currents. Ocean Engineering, 34(11): 1569-1583.

Lick, W., Jin, L. and Gailani, J., 2004. Initiation of movement of quartz particles. Journal of Hydraulic Engineering, 130(8): 755-761.

Liu, J.J. and Yu, G.H., 1995. Study and application on sediment of coastal engineerings. Journal of Nanjing Hydraulic Research Institute, (3): 221-233 (in Chinese with English abstract).

Liu, J.J., 2009. Coastal sediment movement and its application. Ocean Press, Beijing, China. (in Chinese).

Liu, Q., 2007. Analysis of the vertical profile of concentration in sediment-laden flows. International Conference on the Art of Resisting Extreme Natural Forces, pp 355-362.

Longuet-Higgins, M.S. and Stewart, R.W., 1962. Radiation stress and mass transport in gravity waves, with application to 'surf beats'. Journal of Fluid Mechanics, 13(04): 481-504.

Lu, Y.J., Ji, R.Y. and Zuo, L.Q, 2009. Morphodynamic responses to the deep water harbor development in the Caofeidian sea area, China's Bohai Bay. Coastal Engineering, 56(8): 831-843.

Luettich, R.A. and Westerink, J.J., 2004. Formulation and numerical implementation of the 2D/3D ADCIRC finite element model version 44. XX. R. Luettich.

Lundgren, H., 1972. Turbulent currents in the presence of waves. Proc 13th ICCE, Vancouver, ASCE, pp 623-634.

Lofquist, K.E.B., 1986. Drag on naturally rippled beds under oscillatory flows. Misc Paper CERC-86-13, U S Army Corps of Engineers.

Lumborg, U. and Windelin, A., 2003. Hydrography and cohesive sediment modelling: application to the Rømø Dyb tidal area. Journal of Marine Systems, 38(3): 287-303.

Luyten, P., Andreu-Burillo, I., Norro, A., Ponsar, S. and Proctor, R., 2006. A new version of the European public domain code COHERENS. European Operational Oceanography: Present and Future: 474.

Madsen, O.S., 1994. Spectral wave-current bottom boundary layer flows. Coastal Engineering Proceedings, 1(24): 384-398.

Madsen, O.S. and Grant, W.D., 1976. Quantitative description of sediment transport by waves. Coastal Engineering Proceedings, pp 1092-1112.

Madsen, O.S. and Wikramanayake, P.N., 1991. Simple models for turbulent wave-current bottom boundary layer flow, DTIC Document.

Madsen, O.S. and Wood, W., 1993. Sediment transport outside the surf zone. Technical report. US Army Engineer Waterways Experiment Station, Vicksburg, MS.

Manohar, M., 1955. Mechanics of bottom sediment movement due to wave action, DTIC Document.

Maynord, S.T., 1978. Practical Riprap Design, U.S. Army Engineer Waterways Experiment Station, Vicksburg, MS.

McClennen, E.C., 1973. Sands on continental shelf off New Jersey move in response to waves and currents. Maritimes, 14-16.

McLean, S.R., 1992. On the calculation of suspended load for noncohesive sediments. Journal of Geophysical Research: Oceans (1978–2012), 97(C4): 5759-5770.

Mehta, A.J. and Lee, S., 1994. Problems in linking the threshold condition for the transport of cohesionless and cohesive sediment grain. Journal of Coastal Research, 10(1): 170-177.

Meyer-Peter, E. and Müller, R., 1948. Formulas for bed-load transport. Proceedings of the 2nd Meeting of the International Association for Hydraulic Structures Research. International Association of Hydraulic Research Delft, pp 39-64.

Migniot, C., 1968. A study of the physical properties of various very fine sediments and their behaviour under hydrodynamic action. La Houille Blanche, (7): 591-620.

Miller, M., McCave, I.N. and Komar, P.D., 1977. Threshold of sediment motion under unidirectional currents. Sedimentology, 24(4): 507-527.

Ministry of Transport of China, 1998. Code of Hydrology for Sea Harbour, JTJ 213-98. China Communication Press, Beijing (in Chinese).

Mogridge, G.R., Davies, M.H. and Willis, D.H., 1994. Geometry prediction for wave-generated bedforms. Coastal Engineering, 22(3): 255-286.

Myrhaug, D. and Holmedal, L.E., 2007. Mobile layer thickness in sheet flow beneath random waves. Coastal Engineering, 54(8): 577-585.

Myrhaug, D. and Slaattelid, O.H., 1990. A rational approach to wave-current friction coefficients for rough, smooth and transitional turbulent flow. Coastal Engineering, 14(3): 265-293.

Neill, C.R. and Yalin, M.S., 1969. Quantitative definition of beginning of bed movement. Journal of the Hydraulics Division, 95(1): 585-588.

Neumeier, U., Ferrarin, C., Amos, C.L., Umgiesser, G. and Li, M.Z., 2008. Sedtrans05: An improved sediment-transport model for continental shelves and coastal waters with a new algorithm for cohesive sediments. Computers & Geosciences, 34(10): 1223-1242.

Nicholson, J., Broker, I., Roelvink, J.A., Price, D., Tanguy, J.M. and Moreno, L., 1997. Intercomparison of coastal area morphodynamic models. Coastal Engineering, 31(1): 97-123.

Nielsen, P., Svendsen, I.A. and Staub C., 1978. Onshore offshore sediment transport on a beach. Proc 16th Int Conf Coastal Eng, Hamburg, ASCE, pp 1475-1492.

Nielsen, P., 1992. Coastal bottom boundary layers and sediment transport. Advanced Series on Ocean Engineering, 4. World Scientific, Singapore.

Nielsen, P., 1995. Suspended sediment concentration profiles. Appl. Mech. Rev, 48(9): 564-569.

Nielsen, P., van der Wal K.U. and Gillan L., 2002. Vertical fluxes of sediment in oscillatory sheet-flow. Coastal Engineering, 45(1): 61-68.

Nielsen, P. and Callaghan, D.P., 2003. Shear stress and sediment transport calculations for sheet flow under waves. Coastal Engineering, 47(3): 347-354.

Nittrouer, C.A., 1999. STRATAFORM: overview of its design and synthesis of its results. Marine Geology, 154(1): 3-12.

Nittrouer, C.A., Miserocchi, S. and Trincardi, F., 2004. The PASTA project: investigation of Po and Apennine sediment transport and accumulation. Oceanography, 17(4): 46-57.

Normant, C.L., 2000. Three - dimensional modelling of cohesive sediment transport in the Loire estuary. Hydrological processes, 14(13): 2231-2243.

O'Connor, B.A. and Yoo, D., 1988. Mean bed friction of combined wave/current flow. Coastal Engineering, 12(1): 1-21.

O'Donoghue, T. and Wright, S., 2004. Concentrations in oscillatory sheet flow for well sorted and graded sands. Coastal Engineering, 50(3): 117-138.

O'Donoghue, T., Doucette, J.S., Van der Werf, J.J. and Ribberink, J.S., 2006. The dimensions of sand ripples in full-scale oscillatory flows. Coastal Engineering, 53(12): 997-1012.

Pandoe, W.W. and Edge, B.L., 2004. Cohesive sediment transport in the 3D-hydrodynamic-baroclinic circulation model: Study case for idealized tidal inlet. Ocean Engineering, 31(17): 2227-2252.

Parchure, T.M. and Mehta, A.J., 1985. Erosion of soft cohesive sediment deposits. Journal of Hydraulic Engineering, 111(10): 1308-1326.

Partheniades, E., 1965. Erosion and deposition of cohesive soils. Journal of the Hydraulics Division, ASCE, 91(1): 105-139.

Pietrzak, J., Jakobson, J.B., Burchard, H., Jacob Vested, H. and Petersen, O., 2002. A three-dimensional hydrostatic model for coastal and ocean modelling using a generalised topography following co-ordinate system. Ocean Modelling, 4(2): 173-205.

Pinto, L., Fortunato, A.B., Zhang, Y., Oliveira, A. and Sancho, F., 2012. Development and validation of a three-dimensional morphodynamic modelling system for non-cohesive sediments. Ocean Modelling, 57: 1-14.

Ranee, P.J. and Warren, N.F., 1969. The threshold of movement of coarse material in oscillatory flow. Proceedings 11th Conference on Coastal Engineering, ASCE, pp 487-491.

Ravindra Jayaratne, M.P. and Shibayama, T., 2007. Suspended sediment concentration on beaches under three different mechanisms. Coastal Engineering Journal, 49(04): 357-392.

Ribberink, J.S., 1998. Bed-load transport for steady flows and unsteady oscillatory flows. Coastal Engineering, 34(1): 59-82.

Ribberink, J.S. and Al-Salem, A.A., 1995. Sheet flow and suspension of sand in oscillatory boundary layers. Coastal Engineering, 25(3): 205-225.

Ribberink, J.S., van der Werf, J.J., O'Donoghue, T. and Hassan, W.N.M., 2007. Sand motion induced by oscillatory flows: Sheet flow and vortex ripples, Particle-Laden Flow. Springer, The Netherlands.

Richardson, J.F. and Zaki, W.N., 1954. Sedimentation and fluidization: Part I. Transactions of the Institution of Chemical Engineers, 32: 35-53.

Richardson, J.F. and Jeronimo, M.D.S., 1979. Velocity-voidage relations for sedimentation and fluidisation. Chemical Engineering Science, 34(12): 1419-1422.

Righetti, M. and Lucarelli, C., 2007. May the Shields theory be extended to cohesive and adhesive benthic sediments? Journal of Geophysical Research: Oceans (1978–2012), 112(C5): 395-412.

Rigler, J.K. and Collins, M.B., 1983. Initial grain motion under oscillatory flow: a comparison of some threshold criteria. Geo-marine letters, 3(1): 43-48.

Roberts, J., Jepsen, R., Gotthard, D. and Lick, W., 1998. Effects of particle size and bulk density on erosion of quartz particles. Journal of Hydraulic Engineering, 124(12): 1261-1267.

Rodrıguez, D.G., 2009. Wave boundary layer hydrodynamics and cross-shore sediment transport in the surf zone. Massachusetts Institute of Technology.

Roelvink, D., Reniers, A., van Dongeren, A.P., van Thiel, D.V.J., McCall, R. and Lescinski, J., 2009. Modelling storm impacts on beaches, dunes and barrier islands. Coastal Engineering, 56(11): 1133-1152.

Roelvink, D., McCall, R., Mehvar, S., Nederhoff, K. and Dastgheib, A., 2018. Improving predictions of swash dynamics in XBeach: The role of groupiness and incident-band runup. Coastal Engineering, 134: 103-123.

Roelvink, J.A. and Brøker, I., 1993. Cross-shore profile models. Coastal Engineering, 21(1): 163-191.

Roelvink, J.A. and Reniers, A., 2012. A guide to modeling coastal morphology, 12. World Scientific.

Rouse, H., 1937. Modern conceptins of the mechanics of turbulence. Pesticide & Venom Neurotoxicity, 7(3): 1-4.

Ruessink, B.G., Van den Berg, T. and van Rijn, L.C., 2009. Modeling sediment transport beneath skewed asymmetric waves above a plane bed. Journal of Geophysical Research: Oceans (1978–2012), 114: C11021.

Sana, A., Ghumman, A.R. and Tanaka, H., 2007. Modification of the damping function in the k–ε model to analyse oscillatory boundary layers. Ocean Engineering, 34(2): 320-326.

Sato, S., Mimura, N. and Watanabe, A., 1985. Oscillatory boundary layer flow over rippled beds. Coastal Engineering 1984, 2293-2309.

Schoonees, J.S. and Theron, A.K., 1995. Evaluation of 10 cross-shore sediment transport/morphological models. Coastal Engineering, 25(1): 1-41.

Sherwood, C.R., Butman, B., Cacchione, D.A., Drake, D.E., Gross, T.F., Sternberg, R.W., Wilberg, P.L. and Williams, A.J., 1994. Sediment-transport events on the northern California continental shelf during the 1990–1991 STRESS experiment. Continental Shelf Research, 14(10): 1063-1099.

Shields, A., 1936. Anwendung der Ähnlichkeitsmechanik und der Turbulenz Forschung auf die Geschiebe Bewegung, Mitt, der Preuss. Versuchsanst. für Wasserbau und Schiffbau(26).

Sleath, J.F., 1978. Measurements of bed load in oscillatory flow. Journal of the Waterway Port Coastal and Ocean Division, 104(3): 291-307.

Sleath, J.F.A., 1982. The suspension of sand by waves. Journal of Hydraulic Research, 20(5): 439-452.

Sleath, J., 1995. Coastal bottom boundary layers. Applied Mechanics Reviews, 48(9): 589-600.

Smerdon, E.T. and Beasley, R.P., 1961. Critical tractive forces in cohesive soils. Agricultural Engineering, 42(1): 26-29.

Smith, J.D. and McLean, S.R., 1977. Spatially averaged flow over a wavy surface. Journal of Geophysical research, 82(12): 1735-1746.

Soulsby, R., 1997. Dynamics of marine sands: a manual for practical applications. Thomas Telford.

Soulsby, R.L. and Whitehouse, R., 1997. Threshold of sediment motion in coastal environments, Pacific Coasts and Ports' 97: Proceedings of the 13th Australasian Coastal and Ocean Engineering Conference and the 6th Australasian Port and Harbour Conference; Volume 1. Centre for Advanced Engineering, University of Canterbury, pp 149-154.

Sternberg, R.W., 1968. Friction factors in tidal channels with differing bed roughness. Marine Geology, 6(3): 243-260.

Stevens, R.L., 1991. Grain-size distribution of quartz and feldspar extracts and implications for flocculation processes. Geo-Marine Letters, 11(3): 162-165.

Stive, M.J. and De Vriend, H.J., 1994. Shear stresses and mean flow in shoaling and breaking waves. Coastal Engineering 1994, ASCE, pp 594-608.

Styles, R. and Glenn, S.M., 2003. A theoretical investigation of bed-form shapes. Ocean Dynamics, 53(3): 278-287.

Sun, L., Sun, B., Liu, J. and Han, X., 2010. Physical model tests on sudden siltation caused by storm and mitigation measures for fine sand coast in Jingtang port. China Harbour Engineering, S1: 28-31 (in Chinese with English abstract).

Swart, D.H., 1974. Off shore sediment transport and equilibrium beach profiles, Publication 131, Delft Hydraulic Laboratory, Delft, The Netherlands.

Szmytkiewicz, M. et al., 2000. Coastline changes nearby harbour structures: comparative analysis of one-line models versus field data. Coastal Engineering, 40(2): 119-139.

Tanaka, H. and Thu, A., 1994. Full-range equation of friction coefficient and phase difference in a wave-current boundary layer. Coastal Engineering, 22(3): 237-254.

Tanaka, H. and Van To, D., 1995. Initial motion of sediment under waves and wave-current combined motions. Coastal Engineering, 25(3): 153-163.

Tang, C., 1963. Laws of sediment incipient motion. J. Hydraul. Engng, (1): 1-12 (in Chinese with English abstract).

Te Slaa, S., He, Q., van Maren, D.S. and Winterwerp, J.C., 2013. Sedimentation processes in silt-rich sediment systems. Ocean Dynamics, 63(4): 399-421.

Te Slaa, S., van Maren, D.S., He, Q. and Winterwerp, J.C., 2015. Hindered settling of silt. Journal of Hydraulic Engineering, 141(9): 04015020.

Thorne, P.D. and Hanes, D.M., 2002. A review of acoustic measurement of small-scale sediment processes. Continental shelf research, 22(4): 603-632.

Thorne, P.D., Williams, J.J. and Davies, A.G., 2002. Suspended sediments under waves measured in a large-scale flume facility. Journal of Geophysical Research Atmospheres, 107(C8): 3178.

Traykovski, P., P.L. Wiberg and W.R. Geyer, 2007. Observations and modeling of wave-supported sediment gravity flows on the Po prodelta and comparison to prior observations from the Eel shelf. Continental Shelf Research, 27(3): 375-399.

Trowbridge, J. and Madsen, O.S., 1984. Turbulent wave boundary layers: 1. Model formulation and first - order solution. Journal of Geophysical Research: Oceans (1978–2012), 89(C5): 7989-7997.

Trowbridge, J.H. and Kineke, G.C., 1994. Structure and dynamics of fluid muds on the Amazon continental shelf. Journal of Geophysical Research: Oceans, 99(C1): 865-874.

Uittenbogaard, R., Bosboom, J. and van Kessel, T., 2001. Numerical simulation of wave-current driven sand transport. WL| Delft Hydraulics Delft, The Netherlands.

Uittenbogaard, R.E., Bosboom, J. and Kessel, T.V., 2000. Numerical simulation of wave-current driven sand transport: theoretical background of the beta-release of the Point-Sand model, Deltares (WL).

Umeyaina, M., 1992. Vertical Distribution of Suspended Sediment in Uniform Open-Channel Flow. Journal of Hydraulic Engineering, 118(6): 936-941.

Umeyama, M., 2005. Reynolds stresses and velocity distributions in a wave-current coexisting environment. Journal of Waterway, Port, Coastal, and Ocean Engineering, 131(5): 203-212.

Unna, P., 1942. Waves and tidal streams. Nature, 149(3773): 219-220.

van der A, D.A., 2005. 1DV modelling of wave-induced sand transport processes over rippled beds, University of Twente, Enschede, the Netherlands.

van Der Werf, J.J., 2003. A literature review on sand transport under oscillatory flow conditions in the rippled-bed regime, Universty of Twente, Enschede, the Netherlands.

van der Werf, J.J., Ribberink, J.S., O'Donoghue, T. and Doucette, J.S., 2006. Modelling and measurement of sand transport processes over full-scale ripples in oscillatory flow. Coastal Engineering, 53(8): 657-673.

van Doorn, T., 1981. Experimental investigation of near-bottom velocities in water waves without and with a current. Deltares (WL): Delft, M1423, part 1.

van Rijn, L.C., 1984a. Sediment pick-up functions. Journal of Hydraulic Engineering, 110(10): 1494-1502.

van Rijn, L.C., 1984b. Sediment transport, Part II: Suspended load transport. Journal of Hydraulic Engineering, 110(11): 1613-1641.

van Rijn, L.C., 1987. Mathematical modelling of morphological processes in the case of suspended sediment transport.

van Rijn, L.C., 1993. Principles of sediment transport in rivers, estuaries and coastal seas, 1006. Aqua publications Amsterdam.

van Rijn, L.C., 2007a. Unified view of sediment transport by currents and waves. II: Suspended transport. Journal of Hydraulic Engineering, 133(6): 668-689.

van Rijn, L.C., 2007b. Unified view of sediment transport by currents and waves. I: Initiation of motion, bed roughness, and bed-load transport. Journal of Hydraulic Engineering, 133(6): 649-667.

van Rijn, L.C., Nieuwjaar, M.W., van der Kaay, T., Nap, E. and van Kampen, A., 1993. Transport of fine sands by currents and waves. Journal of Waterway, Port, Coastal, and Ocean Engineering, 119(2): 123-143.

Vanoni, V.A., 1964. Measurements of critical shear stress for entraining fine sediments in a boundary layer, California Institute of Technology, Pasadena, California.

Villaret, C., 2010. SISYPHE 6.0 User Manual. Modelisation des Apports Hydriques et Transferts Hydro-Sedimentaires-Laboratoire National d'hydraulique et Environnement.

Villaret, C. and Latteux, B., 1992. Long-term simulation of cohesive sediment bed erosion and deposition by tidal currents computer modelling of seas and coastal regions. Springer, pp 363-378.

Villaret, C., Hervouet, J., Kopmann, R., Merkel, U. and Davies, A.G., 2013. Morphodynamic modeling using the Telemac finite-element system. Computers & Geosciences, 53: 105-113.

Vincent, G.E., 1957. Contribution to the study of sediment transport on a horizontal bed due to wave action. Coastal Engineering Proceedings, 1(6): 326-355.

Waeles, B., Le Hir, P., Lesueur, P. and Delsinne, N., 2007. Modelling sand/mud transport and morphodynamics in the Seine river mouth (France): an attempt using a process-based approach. Hydrobiologia, 588(1): 69-82.

Wai, O., Chen, Y. and Li, Y.S., 2004. A 3-D wave-current driven coastal sediment transport model. Coastal Engineering Journal, 46(04): 385-424.

Wan, Z. and Song, T., 1990. Experimental study on the effect of static pressure on threshold velocity of fine particles. Journal of Sediment Research, (4): 62-69 (in Chinese with English abstract).

Wang, X.H. and Pinardi, N., 2002. Modeling the dynamics of sediment transport and resuspension in the northern Adriatic Sea. Journal of Geophysical Research: Oceans (1978–2012), 107(C12): 18-1-18-23.

Wang, Y.P. et al., 2012. Sediment transport over an accretional intertidal flat with influences of reclamation, Jiangsu coast, China. Marine Geology, 291: 147-161.

Warner, J.C., Sherwood, C.R., Signell, R.P., Harris, C.K. and Arango, H.G., 2008. Development of a three-dimensional, regional, coupled wave, current, and sediment-transport model. Computers & Geosciences, 34(10): 1284-1306.

White, S.J., 1970. Plane Bed Thresholds of Fine Grained Sediments. Nature, 228(5267): 152-3.

Williams, J.J., Bell, P.S., Coats, L.E., Hardcastle, P.J., Humphrey, J.D., Moores, S.P., Thorne, P.D. and Trouw, K., 1998. Evaluation of field equipment used in studies of sediment dynamics, Proudman Oceanogr. Lab., Birkenhead, UK.

Williams, J.J., Rose, C.P., Thorne, P.D., O'Connor, B.A., Humphery, J.D., Hardcastle, P.J., Moores, S.P., Cooke, J.A. and Wilson, D.J., 1999. Field observations and predictions of bed shear stressesand vertical suspended sediment concentration profiles in wave-current conditions. Continental Shelf Research, 19(4): 507-536.

Willis, D.H., 1978. Sediment load under waves and currents. Coastal Engineering Proceedings, ASCE, pp 26-37.

Wilson, K.C., 1989. Friction of wave-induced sheet flow. Coastal Engineering, 12(4): 371-379.

Winterwerp, J.C., 1999. On the dynamics of high-concentrated mud suspensions. Delft University of Technology.

Winterwerp, J.C., 2001. Stratification effects by cohesive and noncohesive sediment. Journal of Geophysical Research: Oceans (1978–2012), 106(C10): 22559-22574.

Winterwerp, J.C. and Uittenbogaard, R.E., 1997. Sediment transport and fluid mud flow: physical mud properties and parameterization of vertical transport processes, Deltares (WL).

Winterwerp, J.C. and Van Kesteren, W.G., 2004. Introduction to the physics of cohesive sediment dynamics in the marine environment. Elsevier.

Winyu, R. and Shibayama, T., 1995. Suspended sediment concentration profiles under non-breaking and breaking waves. Coastal Engineering 1994, 2813-2827.

Wright, L.D., Boon, J.D., Kim, S.C. and List, J.H., 1991. Modes of cross-shore sediment transport on the shoreface of the Middle Atlantic Bight. Marine Geology, 96(1): 19-51.

Wu, W., 2008. Computational river dynamics. CRC Press.

Wu, W., Sanchez, A. and Zhang, M., 2010. An implicit 2-D depth-averaged finite-volume model of flow and sediment transport in coastal waters, DTIC Document.

Xia, Y., Xu, H., Chen, Z., Wu, D. and Zhang, S., 2011. Experimental study on suspended sediment concentration and its vertical distribution under spilling breaking wave actions in silty coast. China Ocean Engineering, 25(4): 565-575.

Xiao, H., Cao, Z. and Zhao, Q., 2009. Experimental study on incipient motion of coherent silt under wave and flow action. Journal of Sediment Research, 16(3): 75-80 (in Chinese with English abstract).

Yalin, M.S., 1963. An expression for bed-load transportation. J. Hydraul. Div. Am. Soc. Civ. Eng, 89(3): 221-250.

Yalin, M.S. and Karahan, E., 1979. Inception of sediment transport. Journal of the Hydraulics Division, 105(11): 1433-1443.

Yang, H. and Hou, Z.Q., 2004. Study on siltation in the outer channel of Huanhua Harbor. Journal of Waterway and Harbor, 25: 59-63 (in Chinese with English abstract).

Yang, M. and Wang, G., 1995. The incipient motion formulas for cohesive fine sediments. Journal of Basic Science and Engineering, 3(1): 99-109 (in Chinese with English abstract).

Yang, S., Tan, S., Lim, S. and Zhang, S., 2006. Velocity distribution in combined wave–current flows. Advances in water resources, 29(8): 1196-1208 (in Chinese with English abstract).

Yao, P., Su M., Wang, Z.B., van Rijn, L.C., Zhang, C.K., Chen, Y.P., Stive, M.J., 2015. Experiment inspired numerical modeling of sediment concentration over sand-silt mixtures. Coastal Engineering, 105: 75-89.

Ye, Q., 2006. Modelling of Cohesive Sediment Transportation, Deposition and Resuspension in the Haringvliet Mouth. Diss., UNESCO-IHE, the Netherlands.

You, Z., 1994. Eddy viscosities and velocities in combined wave-current flows. Ocean Engineering, 21(1): 81-97.

You, Z., 1998. Initial motion of sediment in oscillatory flow. Journal of Waterway, Port, Coastal, and Ocean Engineering, 124(2): 68-72.

You, Z., Wilkinson, D.L. and Nielsen, P., 1991. Velocity distributions of waves and currents in the combined flow. Coastal Engineering, 15(5): 525-543.

Zhang, C., Zheng J.H., Wang, Y.G., Zhang, M.T., Jeng, D.S. and Zhang, J.S., 2011. A process-based model for sediment transport under various wave and current conditions. International Journal of Sediment Research, 26(4): 498-512.

Zhang, C., Zheng, J.H., Wang, Y.G. and Demirbilek, Z., 2011. Modeling wave–current bottom boundary layers beneath shoaling and breaking waves. Geo-Marine Letters, 31(3): 189-201.

Zhang, Q., Bing, Y. and Onyx, W.H., 2009. Fine sediment carrying capacity of combined wave and current flows. International Journal of Sediment Research, 24(4): 425-438.

Zhao, Q. and Han, H., 2007. Siltation mechanisms of Huanghua Port and 3 D characteristics of nearshore suspended sediment concentration under waves. Journal of Waterway and Harbor, 28(2): 77-80 (in Chinese with English abstract).

Zhao, Z., Lian, J. and Shi, J.Z., 2006. Interactions among waves, current, and mud: numerical and laboratory studies. Advances in water resources, 29(11): 1731-1744 (in Chinese with English abstract).

Zheng, J., Li, R.J., Feng, Q. and Lu S.S., 2013. Vertical profiles of fluid velocity and suspended sediment concentration in nearshore. International Journal of Sediment Research, 28(3): 406–412.

Zhou, Y. and Ju, L., 2007. Test study on the characteristics of a wind-flow-wave flume and sediment movement under wave conditions, Nanjing hydraulic research institute, Nanjing (in Chinese with English abstract).

Zhou, Y., Chen, Y. and Ma, Q., 2001. Threshold of sediment movement in different wave boundary layers. China Ocean Engineering, 15(4): 509-520.

Zuo, L., Lu, Y., Wang, Y. and Liu, H., 2014. Field observation and analysis of wave-current-sediment movement in Caofeidian Sea area in the Bohai Bay, China. China Ocean Engineering, 28 (3): 331-348.

Zuo, L., Roelvink, D., Lu, Y. and Li, S., 2017. On incipient motion of silt-sand under combined action of waves and currents. Applied Ocean Research, 69: 116-125.

Zyserman, J.A. and Fredsøe, J., 1994. Data analysis of bed concentration of suspended sediment. Journal of Hydraulic Engineering, 120(9): 1021-1042.

Acknowledgements

This PhD thesis is the result of a joint research project of The Netherlands Organisation for Scientific Research (NWO) and National Natural Science Foundation of China (NSFC). It was indeed a never-forgettable experience in my life to do the PhD research in the past 4 years. It provided me a good opportunity to do research in an international academic environment. It is not only an excellent academic experience, but also consists different ways of thinking, life styles, cultures and more dreams. During this period, I had to constantly travel between Delft and Nanjing for my PhD research and consultant projects, which was really a tough time for me. However, all the efforts paid off when the thesis was completed. I would like to express my sincere appreciation to all organizations and people that stimulated me to conduct this research.

My first thank belongs to promoter, Professor Dano Roelvink. My appreciation and gratitude go much further than what I can write here. His expertise, passion and thoughtful guidance always offer me great ideas and insight. Many thanks for his knowledge, encouragement, patience and numerous time spent in keeping me in the right track. Even when he was busy, a short talk can still show me light in the darkness.

Special thanks are given to Professor Yongjun Lu. It is his inspiring and unwavering support that guided me in the field of sediment research. I am indebted to his care in my career life.

Colleagues in CEPD of IHE, Rosh, Mick, Ali, Johan, Trang, Duoc, Hao, Tham, Jakia, Seyedabdolhossein, Janaka, Hesham, Jentsje, Shah, Fernanda, Leo, Shah, Jeewa, Uwe, etc., are acknowledged for nice conversations and help.

As I travelled frequently between the Netherlands and China, I appreciate Jolanda for her nice and patient arrangement. And I want to thank Zhaoheng Liu, Rong Chen, Feng Sun, Lanping Xiao and Tiantao Zhou for their help in China side related to travelling affairs.

Thank Anique and Floor for their help in arranging my registration and defence. Thank Paula for the help in the format checking and thesis publishment. I thank Tonneke, Sylvia, Marielle, Niamh, etc. for their help during my stay in Delft.

Appreciations are hereby extended to Yaping Wang, Jianhua Gao, Benwei Shi, as well as Master students Yong Shi, Yiyi Zhang, Mingliang Li and Yingfei Li from Nanjing University for their help in the field observation process. Our gratitude also belongs to Binfeng Zhu, Huaixiang Liu, Tingjie Huang and Weihao Huang from NHRI for their help with field observation and indoor instrument calibration and data processing. Thanks a lot Hesen Cheng, Mingcheng Wei and Jie Shen from NHRI for their help in the density measurement.

I'd like to give my special thanks to Qinghua Ye and Taoping Wan, for their help during my stay in Delft. Helps from Leicheng Guo, Yuqing Lin, Yuanyang Wan, Chunqing Wang, Sien Liu, Zhi Yang, Sida Liu, Xiuhan Chen, Xuhui Chen, Quan Pan, Danqian Shen, Jingyi Chen, Ao Chu, Taotao Zeng, Lingna Wei, Yingming Hu, Xuan Zhu, Victor, Polpat, Poldul, Saowanit, Aries, Maria, Ha, Than, Nguyen, Clara, Lan, Hieu, Vitali, etc. are also cherished during my life in Delft.

I'd also extend my sincere thanks go to my colleagues at NHRI I am grateful to Yunfeng Xia, Junning Pan, Guobin Li, Jianzhong Wang, Yan Lu, Zhili Wang, Shouqian Li, Huaixiang Liu, Qingzhi Hou, Tingjie Huang, Yi Liu, Chi Ma, Xianglong Wei etc. for their encouragement and support in my study.

I would like to thank the committee for their willingness to evaluate and their helpful comments on this thesis.

Numerous friends met in Delft and in Nanjing, beyond a name list I can make, are thanked for your accompany and help. Wish you all the best wherever you are.

In addition, I want to thank my parents and parents in law, for their constant and unconditional support. I own my most sincere thank you to my wife Tingting Wang. You deal with the family matters and brought up our daughter when I was abroad. The final thank goes to my daughter, Xiaoyou Zuo. I left you to the Netherlands when you were three. Four years past and you have stepped into primary school from kindergarten as my research started from a scratch to the final report. You enriched the meaning of my life and enlightened my world. Thank all for your love and support; it is you that I owe everything.

Liqin Zuo
Delft, May 2018

About the Author

Liqin Zuo was born on 29 September, 1980 (based on Chinese lunar calendar) in Juye, Shandong Province, northern China. He studied Hydrology and Water Resource for bachelor degree in Hohai University from 1999 to 2003. He studied Harbor, Coastal and Offshore Engineering in Nanjing Hydraulic Research Institute (NHRI) from September 2003, and got his master degree in 2006. He has been working in NHRI since July 2006. He conducted his part-time PhD research from 2008 to 2015 in NHRI, working on the sediment transport and river bed evolution in upstream and downstream Yangtze River of the Three Gorges Project. In 2014, he was promoted as senior engineer. He participated the joint research project of NWO-NSFC between China and the Netherlands, for which he started his PhD study in the Netherlands. He arrived at Delft on 10 July, 2014 and since then has been working on the topic of sediment alluvial process in wave-current bottom boundary layer.

Exposure

Publications
Selected papers on peer-reviewed journals

Liqin Zuo, Dano Roelvink, Yongjun Lu, and Shouqian Li., 2017. On incipient motion of silt-sand under combined action of waves and currents. Applied Ocean Research, 69: 116-125.

Liqin Zuo, Yongjun Lu, Yaping Wang, and Huaixiang Liu, 2014. Field observation and analysis of wave-current-sediment movement in the Caofeidian sea area in the Bohai Bay, China. China Ocean Engineering. 28 (3): 331 – 348.

Liqin Zuo, Dano Roelvink, Yongjun Lu, Hao Wang. Modelling and analysis on high sediment concentration layer of fine sediments under wave-dominated conditions. (under reviewed)

Liqin Zuo, Dano Roelvink, Yongjun Lu. The mean suspended sediment concentration of silty sediment under wave-dominated conditions. (submitting)

Yongjun Lu, Shouqian Li, **Liqin Zuo**, Huaixiang Liu, and J. A. Roelvink, 2015. Advances in sediment transport under combined action of waves and currents. International Journal of Sediment Research, 30 (4): 351-360.

Yongjun Lu, Rongyao Ji, and **Liqin Zuo**, 2009. Morphodynamic responses to the deep water harbor development in the Caofeidian sea area, China's Bohai Bay. Coastal Engineering, 56 (8): 831-843.

Yongjun Lu, **Liqin Zuo**, Rongyao Ji, Xiao Xu, and Jianwei Huang, 2008. Effect of development of Caofeidian harbor area in Bohai Bay on hydrodynamic sediment environment. China Ocean Engineering, 22(1) 97-112.

Yongjun Lu, **Liqin Zuo**, Xuejun Shao, Hongchuan Wang, and Haolin Li, 2005. A 2D mathematical model for sediment transport by waves and tidal currents. China Ocean Engineering, 19(4): 571-586.

Liqin Zuo, Yongjun Lu, and Rongyao Ji, 2014. Back silting of the mouth bar channel in southern Zhejiang muddy coastal estuaries. Journal of Basic Science and Engineering, 22(1): 88-105. (in Chinese with English abstract)

Liqin Zuo, Rongyao Ji, and Yongjun Lu, 2012. Case study of siltation in channel-mouth bar in offshore barrier-lagoon coast. Advances in Water Science, 23(1): 87-95. (in Chinese with English abstract)

Book Chapter

Yongjun Lu, **Liqin Zuo**, Rongyao Ji, et al., Chapter 9 Sediment transport in the Caofeidian Coastal Area of Bohai Bay, China. ——Abdul A. Khan and Weiming Wu, 2013, "Sediment transport—Monitoring, Modeling and Management", NOVA Science Publishers, USA.

Conference participation

Liqin Zuo, Dano Roelvink and Yongjun Lu, 2015. Advances in numerical simulation for sediment transport under co-acction of waves and currents. 36th IAHR, The Hague, the Netherlands.

Liqin Zuo, Yongjun Lu, 2013. Back silting feature in the mouth bar channel in Aojiang estuary. 35th IAHR, Chengdu, China.

Liqin Zuo, Yongjun Lu and Rongyao Ji, 2012. Study on siltation in Laolonggou mouth bar channel in Caofeidian sea area of Bohai bay, China. 4th International Conference on Estuaries and Coasts (ICEC 2012), Hanoi, Vietnam.

Netherlands Research School for the
Socio-Economic and Natural Sciences of the Environment

D I P L O M A

For specialised PhD training

The Netherlands Research School for the
Socio-Economic and Natural Sciences of the Environment
(SENSE) declares that

Liqin Zuo

born on 29 September 1980 in Juye city, Shandong, China

has successfully fulfilled all requirements of the
Educational Programme of SENSE.

Delft, 4 June 2018

the Chairman of the SENSE board

Prof. dr. Huub Rijnaarts

the SENSE Director of Education

Dr. Ad van Dommelen

K O N I N K L I J K E   N E D E R L A N D S E
A K A D E M I E   V A N   W E T E N S C H A P P E N

The SENSE Research School declares that Mr Liqin Zuo has successfully fulfilled all
requirements of the Educational PhD Programme of SENSE with a
work load of 32.6 EC, including the following activities:

<u>SENSE PhD Courses</u>

o Environmental research in context (2014)
o Writing week (2017)
o Research in context activity: co-author of (awarded) research proposal on: 'The tidal flat-
  channel morphodynamic process of silty-muddy coast and its response to large scale
  contiguous reclamation' (research period 2016-2020) AND
  'Co-organizing workshop for cooperation between IHE Delft and NHRI (Nanjing Hydraulic
  Research Institute, China) on: 'The tidal flat-channel morphodynamic process of silty-
  muddy coast and its response to large scale contiguous reclamation' (2017)

<u>Other PhD and Advanced MSc Courses</u>

o Coastal morphodynamics, UNESCO-IHE (2015)
o Waves in oceanic and coastal waters, UNESCO-IHE (2015)
o Art of presenting science, UNESCO-IHE (2017)

<u>External training at a foreign research institute</u>

o Morphological modelling using Delft3D, UNESCO-IHE (2014)
o Delft Software day, Deltares, The Netherlands (2016)

<u>Management and Didactic Skills Training</u>

o Co-organiser Fourth Workshop on 'Sediment alluvial process in the wave-current
  boundary layer', Delft (2014)
o Guide to students for field trip 'Hydraulic engineering works in South-West Netherlands'
  (2017)

<u>Selection of Oral Presentations</u>

o *Advance in sediment transport under co-action of waves and currents.* PhD symposium
  of UNESCO-IHE, 29-30 September 2014, Delft, The Netherlands
o *Advances in numerical simulations for sediment transport under co-action of waves and
  currents.* E-proceedings of the 36th IAHR World Congress, 28 June- 03 July 2015, The
  Hague, The Netherlands
o *Flow-sediment dynamics of the fine sediment in the wave-current bottom boundary
  layer: modelling and parameterization.* Workshop on the joint research project, 10
  October 2017, Nanjing, China

SENSE Coordinator PhD Education

Dr. Peter Vermeulen

T - #0415 - 101024 - C204 - 240/170/11 - PB - 9781138334687 - Gloss Lamination